Earth's Magnetism in the Age of Sail

Earth's Magnetism in the Age of Sail

A. R. T. Jonkers

The Johns Hopkins University Press
Baltimore and London

Publication of this book was facilitated through a Research publication
grant awarded by the Royal Society of London.

The Johns Hopkins University Press
2715 North Charles Street
Baltimore, Maryland 21218-4363
www.press.jhu.edu

Library of Congress Cataloging-in-Publication Data

Jonkers, A.R.T., 1967–
  Earth's magnetism in the age of sail / A.R.T. Jonkers.
        p.   cm.
  Includes bibliographical references and index.
  ISBN 0-8018-7132-8 (hardcover : alk. paper)
  1. Geomagnetism—History.   2. Compass—History.   3. Navigation—
History.   I. Title.
  QC813 .J66 2003
  538'.7'09—dc21   2002005384

A catalog record for this book is available from the British Library.

Contents

Figures and Tables

TABLES

Acknowledgments

Many people have been involved in this project; some in a professional capacity, others devoting much of their spare time to help me with translation, proofreading, and providing valuable insights and suggestions. The four people most closely associated with the research were Karel Davids, David Gubbins, Andrew Jackson, and Anne Murray, to whom I am deeply indebted for guidance, input, support, and companionship. Various individuals in England have likewise contributed to no mean extent to the completion of the work. Among those (formerly or presently) working at the Department of Earth Sciences at the University of Leeds, I would like to mention Gianluca "cow-boy" Badalini, Nick Barber, Peter Kelly, Vincent Lesur, Jeff Love, Jürgen "Locko" Neuberg, Aoife O'Mongain, Christine Thomas, and Matthew Walker. Of people encountered at other English institutions, I wish to thank Simon Bailey, David Barraclough, Andrew Davis, Mark Greengrass, Willem Hackmann, Jack Jacobs, Nick Kusznir, Chris Paul, Stephen Pumfrey, Kathy Whaler, and Glyndwr Williams. In the Netherlands, I am grateful to Jaap Bruijn, Jent Bijlsma, A. F. van Engelen, Koen Goudriaan, Arno Grünewald, René Isarin, Eric Jorink, Frits Koek, Cor Langereis, Sandra Langereis, Alan Lemmers, Willem Mörzer Bruyns, Agnes Rutgers, Caroline Schuurmans-Hagedoorn, Marike Verschoor, Marieke van der Werf, and Siebren van der Werf. In Denmark, I salute Eigil Friis-Christensen, Paul Frick, Knud Frydendahl, Erik Gøbel, Claus Jensen, Camilla Rygaard-Hjalstedt, and Søren Thirslund. Furthermore, in the United States I am indebted to Jeremy Bloxham and Gregory H. Good, as well as to Bob Brugger, Melody Herr, Marie Blanchard, Tom Roche, and others associated with the Johns Hopkins University Press, who have helped immensely in turning my dissertation into a book.

I also extend many thanks to all members of staff at museums, libraries, archives, and other institutions everywhere, for providing information on, and access to the historical material. Despite the contributions of those mentioned above, I naturally take full responsibility for any remaining errors in the text.

Funding for the research was provided by the Netherlands Organization for Scientific Research (NWO), and the National Environment Research Council (NERC) grants GR9/01848 and GR3/10581.

Note on Spelling and Other Conventions

Quotations from primary sources have been normalized to the present standard in punctuation and usage of upper and lower case. The only modernization of spelling of these texts is the replacement of Old English *y* (in *ye* and *y.t*, for instance) by *th* (to *the* and *th.t*).

Personal names most often follow their owners' spelling. In some instances, their Latinized pen name is adhered to (for instance, "Mercator" instead of "Kremer"). Geographical locations are referred to in modern spelling except in a direct quotation from a primary source; their designation may use historical names (for instance, "Ceylon" instead of "Sri Lanka").

Dates are represented in Italian style: DD-MM-YYYY; a given range of dates (or other quantities) is implicitly inclusive. Angles and measures of arc are often expressed in degrees and minutes (historical sources), less frequently in decimal degrees (modern statistical analysis). If not made explicit, the distinction is usually apparent in the notation: $12°34'$ equals 12 degrees and 34 (sexagesimal) minutes, whereas 12.34 degrees denotes 12 degrees and 34/100 (decimal).

Abbreviations

Add	Additional (mss, BL)
ADM	Admiralties
Adv	Advocates (mss, NLS)
AGI	Archivo General de Indias, Seville, Spain (India Archive)
AK	Asiatisk Kompagni (Danish East India Company)
ANP	Archives Nationales, Paris, France (French National Archives)
ARA	Algemeen Rijks Archief, The Hague (Dutch National State Archive)
ATL/Atl	Atlantic
Bib	*Bibliotheek* (library)
BL	British Library, London, England
BNP	Bibliothèque Nationale, Paris, France (French National Library)
BODL	Bodleian Library, University of Oxford, England
BOL	Board of Longitude
BPL	Bibliothecae Publicae Latini (Latin mss, UBL)
CDI	Compagnie des Indes (French East India Company)
CMB	Core-Mantle Boundary
CP/Cl.P.	Classified Papers (mss, RS)
CUL	Cambridge University Library, Cambridge, England
D	magnetic declination
DUC	papers of Sir John Thomas Duckworth (mss, NMM)
EI	East Indies, East India
EIC	(English) East India Company
E-W	east-west
FAD	French Admiralty (French Navy)
FR	*Français* (French mss, BNP)
FRS	Fellow of the Royal Society (of London)
FSM	Fries Scheepvaart Museum, Sneek, The Netherlands (Frisian Shipping Museum)

GAA	Gemeente Archief Amsterdam, The Netherlands (Municipal Archives)
GAD	Gemeente Archief Delft, The Netherlands (Municipal Archives)
GADord	Gemeente Archief Dordrecht, The Netherlands (Municipal Archives)
GAR	Gemeente Archief Rotterdam, The Netherlands (Municipal Archives)
GAS	Gemeente Archief Schiedam, The Netherlands (Municipal Archives)
GK	Guinesk Kompagni (Danish Guinea Company)
HBC	Hudson's Bay Company
HHRL	Hoogheemraadschap Rijnland, Leyden, The Netherlands (Dike-reeve High Office)
HMN	papers of Commander John Hamilton (mss, NMM)
I	magnetic inclination
im	*imagen* (image, AGI)
IOBL	India Office, British Library (BL)
KGH	Kongelige Grønlandse Handelskompagni (Royal Greenland Trade Company)
KNAG	Koninklijk Nederlands Aardrijkskundig Genootschap, Utrecht, The Netherlands (Royal Dutch Geographical Society)
KNMI	Koninklijk Nederlands Meteorologisch Instituut, De Bilt, The Netherlands (Royal Dutch Meteorological Institute)
LOG	logbooks (mss, NMM)
LP	Letters and Papers (mss, RS)
MAR	Maritime (mss, ANP/IOBL)
MCC	Middelburgsche Commercie Compagnie (Dutch slave trading company)
MER	merchant marine
MM	Miscellaneous Manuscripts (mss, RS)
MMPH	Maritiem Museum "Prins Hendrik", Rotterdam, The Netherlands
MSS	manuscripts
NAF	*Nouvelles Acquisitions Françaises* (New French Acquisitions, mss, BNP)
NAV	navy/navies
NE	northeast(ing)
NLS	National Library of Scotland, Edinburgh, Scotland

NMM	National Maritime Museum, Greenwich (London), England
N-S	north-south
NSM	Nationaal Historisch Scheepvaart Museum, Amsterdam, The Netherlands (Dutch National Shipping Museum)
NVP	papers relating to navigational shipping practice (mss, NMM)
NVT	papers relating to navigational theory (mss, NMM)
NW	northwest(ing)
PA	*particulier archief* (private archive)
preVOC	Dutch EICs prior to the VOC (1597–1602)
PRO	Public Record Office, Kew (London), England
RAK	Rigsarkivet, Copenhagen, Denmark (Danish National Archive)
RAZ	Rijksarchief Zeeland, Middelburg, The Netherlands (provincial archive)
RGO	Royal Greenwich Observatory, Greenwich (London), England
RN	Royal Navy (England)
RS	Royal Society, London, England
SE	Sø-Etaten (Danish Navy)
SG	Staten-Generaal (Dutch Estates General, mss, ARA)
SLA	slave trade
THO	papers of Vice-Admiral Sir Charles Thompson (mss, NMM)
TID	Tiddeman papers (mss, NMM)
UBL	University of Leyden Library, The Netherlands
VA	Vejledene Arkivregistraturet (Guide to the Archives, RAK)
VOC	Verenigde Oostindische Compagnie (Dutch East India Company)
WEL	Wellcome papers (mss, NMM)
WHA	whaling

Earth's Magnetism in the Age of Sail

Merging Geomagnetism and History

In 1904 a young American named Andrew Ellicott Douglass started to collect tree specimens. He was not seeking a pastime to fill his hours of leisure; his motivation was purely professional. Yet he was not employed by any forestry department or timber company, and he was neither a gardener nor a botanist. For decades he continued to amass chunks of wood, all because of a lingering suspicion that a tree's bark was shielding more than sap and cellulose. He was not interested in termites, or fungal parasites, or extracting new medicine from plants. Douglass was an astronomer, and he was searching for evidence of sunspots.

Solar activity is governed by eleven-year cycles, featuring fairly predictable changes in a number of phenomena. Some of these can affect conditions on Earth as well: magnetic storms, for instance, interfere with radio communications, and cause the spectacular display of the northern lights in high latitudes. The Sun's strong magnetic fields also generate turbulent patches of lower than average temperature on its surface. These appear as clusters of dark spots, which reduce solar brightness in proportion to their size. Changes in the total amount of emitted energy in turn influence weather and climate on our planet. In addition, some species of trees are quite sensitive to variability in annual temperature and rainfall. Their growth is stimulated during warm or humid years, and reduced in cold or dry ones. Tree ring formation thus creates a faithful and lasting record of the average prevailing weather conditions in the past.

It should therefore come as no surprise that Douglass was ultimately successful in his quest. But he is mentioned here because he made an even more remarkable discovery in the process. Instead of confining his samples to living trees in nearby woodlands, he added more ancient material, from old buildings, archaeological finds, and pieces of wood untouched by man, preserved in the soil. A sufficiently long matching succession of wide and narrow rings witnessed in separate pieces constitutes an overlap in time. Bit by bit, a coherent picture slowly grew; by the late twentieth century, researchers had for some regions extended the sequence several millennia back in time. The scientific discipline of *dendrochronology* nowadays provides the archaeologist with a dating tool precise to the year, while *dendroclimatology* offers insights into local and global climate change. A wealth of data has suddenly become available to many, after lying dormant in forests for centuries.

These sequences of tree rings can be interpreted as a veiled layer of information. The exposure of such a "hidden level" relies on the recognition of the underlying pattern, through careful study of a sufficiently large sample. Common traits can be identified only when signal dominates noise. In this respect, numerous problems affect tree ring analysis: some species and specimens of trees are simply unsuitable. In others, radial growth is correlated with moisture in certain latitudes, whereas elsewhere the temperature is decisive. Abnormal circumstances can furthermore generate false or irregular rings, or prevent one from forming altogether. Yet despite these potential difficulties, Douglass eventually succeeded in exploiting this hitherto untapped resource. None before him had done so, even though the same subtle clues had been accessible to biologists for ages. But the astronomer had a decided advantage in that he knew what he was looking for, based on particular knowledge from his own field of investigation. The development of dendrochronology therefore provides an example of the potential rewards of an interdisciplinary approach.

Regrettably, such communication is frequently merely one-way; the benefits tend to primarily befall only one of the disciplines involved. One branch of inquiry is then made subservient to the goals of another, without the recipient being able to reciprocate in equal measure, or at all. Ideally, interdisciplinary investigations should attempt to satisfy both parties involved, but in practice this appears difficult to achieve. Researchers tend to be specialists in their chosen fields, their interests only including those other areas of study that can directly affect their own. This is not harmful in itself, but it may impede the exchange of ideas in a wider context. Moreover, research methodology, terminology, and protocol may differ substantially from one discipline to the next. Consequently, many opportunities for intellectual cross-fertilization are lost. Few modern mathematicians will, for example, appreciate the part played by London's Gresham College, during the early seventeenth century, in the dissemination of the concept of the natural logarithm. Few historians will, on the other hand, be aware of the full implications of the power of logarithms in facilitating various calculations. Yet a comprehensive understanding of its development and ramifications would require both aspects to be taken into account. A historian of science will thus need to have some grounding in the matters scrutinized by his protagonists, not only to be able to follow their reasoning, but also in order to understand where, when, and how their interpretations differed from the current one.

Joining Forces

The research described in the following chapters intends to cross the gap between two very dissimilar disciplines: geomagnetism and history. It is a case study in interdisciplinarity in the true sense of the word, since it belongs fully to neither one nor to the other, but tries instead to perform a balancing act between the two. Central to this effort is the Earth's magnetic field. At its deep origin in the Earth's core, liquid iron is driven by intense heat and rotation to maintain a system of convection. This slow circulation generates a magnetic field strong enough to align a compass needle at the surface. To this day, no comprehensive theory exists that can adequately explain and predict the field's changing appearance on different time scales. Part of the problem is the relatively small window of analysis, given that particular features take several centuries to grow, migrate across the globe, and dwindle again. The science of geomagnetism is simply too young to have established a sufficiently long observational tradition by itself. Other, more ancient sources of information have to be included for meaningful analysis of intermediate and long-term changes. This is where history can prove a valuable ally.

Picture for a moment a magnetic measurement as a small source of light. Most of historical time then remains in utter darkness, with tiny twinkles here and there, lighting up for an instant at a site where a compass reading was recorded. Only some major European cities would then be able to maintain a more or less constant flickering through the ages; the other continents would remain almost entirely dark. But from the sixteenth century, the oceans would start to display a quite remarkable spectacle, with ever brighter corridors across the Atlantic and the Indian Oceans, and individual traces snaking along coastlines, circling islands, and occasionally crossing the Pacific.

In the seventeenth and eighteenth centuries, the geomagnetic field was even less well understood than at present. At sea it was both friend and foe, guiding mariners, but also confounding them by constantly changing appearance over the years. Therefore, a vigil had to be kept through frequent observation, stored in the ship's navigational logbook, and allowance made for needle deflection from true north in steering, dead-reckoning, and charting. As the busi-

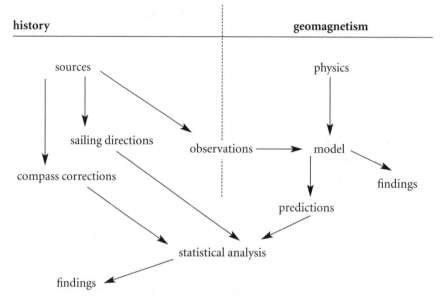

Fig. I.1. Research flow chart: historical sources provide the observations for the geophysical model; the latter can be statistically compared with historical predictions (in sailing directions and logbooks) of magnetic declination; new information benefits befall both disciplines.

ness of seafaring created a wide network of shipping lanes, compass behavior along vast ocean traverses and distant shores was thus recorded. Meanwhile, back at home, hydrographers compiled tables of measurements and updated sailing directions to reflect the field's alterations. Maritime powers thereby accumulated a large body of data, which covered a significant part of the world's oceans. These provide rich pickings for geomagnetists today, spanning an era before land-based observatories, regular naval surveys, and global satellite coverage became commonplace. Together, they can spawn a geomagnetic model unprecedented in scope, accuracy, and detail.

But as thrilling as this project must have appeared to a handful of geophysical specialists, precious few historians will have lost any sleep over it. Merely collecting numerical data numbs the brain and does nothing to enhance our understanding of the past. A physical description of the field, moreover, requires abstract mathematics and powerful computers to explore the darkest corners of multidimensional "solution space"—scary and seemingly incomprehensible to most people dedicated to old books and manuscripts. There would have to be a substantial reward for a historian to venture into this realm of calculus and fleeting flux. And that is exactly what this effort sets out to discover, because the story does not end with the reconstruction of the field in former times. The groundbreaking part of this research lies in feeding the geomagnetic findings back into the historical domain. A physically constrained model of the changing magnetic field can provide an independent touchstone with which to compare navigational practice in the age of sail. How well different nations, maritime organizations, and individuals were able to deal with this unpredictable phenomenon can then be quantified for the very first time. The interdisciplinary approach is therefore based on true reciprocity.

Very little existing research into seventeenth- and eighteenth-century navigation has up to now been quantitative in nature. Consequently, very few statements have been made regarding differences in accuracy attained during measurement of, and correction for, the straying magnetic needle, both in navigation and hydrography. Ships' logbooks and sailing directions offer a wealth of material, eminently suited for numerical investigation of various subsets of data. The same processing effort that made available the observations for the field model also engendered the largest ever compilation of dead-reckoning and steering estimates of local magnetic declination (known as *compass allowance, declination allowance,* or simply *allowance*). Comparison of these two datasets has never before been attempted on remotely the same scale. Moreover, the statistical discipline requires a certain minimum sample size to allow a reasonable level of confidence to be associated with any finding. Many of the conclusions drawn in the final chapter are therefore completely novel, since these conditions have not previously been met in maritime studies.

Bulk analysis of historical sources not only offers a better model to the geophysicist, but also constitutes a powerful tool for historians to examine the subject. Tens of thousands of scattered notes, which by themselves carry little explanatory value beyond their immediate context, conceal both a planetwide geophysical picture *and* a distinctive imprint of maritime expertise. On the most fundamental level, this book will try to establish to what extent these hidden layers of information can be exploited by both disciplines concerned.

The main reason for choosing geomagnetism for this experiment lies in the fact that the Earth's changing field has left an indelible mark on seafaring, hydrography, and natural philosophy, a record which has so far been severely underrated. *Magnetic declination,* the angle between geographic and magnetic north, was not necessarily an impediment while traversing the oceans. Navigators and hydrographers associated knowledge of the observed spatial distribution of needle behavior with specific geographical landmarks. Hence, a magnetic reading could serve to render a rough positional fix. More sophisticated schemes equated a certain distance traveled east or west with a certain change in magnetic declination. Mariners resorted to these measures because they lacked a practical method with which to observe longitude at sea. Furthermore, the various financial rewards promised to the one who would solve this navigational problem generated a host of "longitude solutions" that built upon geomagnetic hypotheses. These postulates invariably supposed some kind of global regularity in the distribution of field lines, which would allow an easily observed "magnetic coordinate" to replace the imprecise longitudinal estimates used in dead-reckoning. Some inventors were purely intent on satisfying a practical need; others spent years building intricate conjectures of causal mechanisms, deemed active in the heavens, the planet's crust, or its deep interior, depending on where the field was thought to originate. The matter occupied the minds of mariners, natural philosophers, compass makers, physicians, mathematical practitioners, hydrographers, and sundry cranks of diverse plumage. Unfortunately, all protagonists of such theories have underestimated the field's complexity. Their notions do, however, testify to a remarkable zeal in the pursuit of insight, fortune, and glory.

In addition, these ideas not only reflected and affected scientific progress in general, but also displayed intrinsic dynamics in the spread and succession of four phases of development. The presently available reconstruction of the Earth's magnetic field in former times now enables us to visualize the geomagnetic circumstances at the time when various individuals formed their theories. This approach can yield clues as to whether, and to what extent, actual observations (*empirical evidence*) may have driven this pursuit of knowledge. The propositions also distinctly bear the hallmarks of their intellectual environment, shaped as they were by competing natural philosophies. This interaction enables these studies to cast light on wider-ranging contemporary issues, such as the debate between *geocentrists* and *heliocentrists*, on whether the Earth or the Sun was the center of the cosmos, and discussions concerning the nature of the inner Earth. The formation of geomagnetic concepts moreover display a coupling between a growing number of available data on the one hand, and a shift from *inductive*, causal hypotheses toward mostly *deductive* descriptions. The genesis, transmission, and competition of these ideas constitutes a rich field of study, one that deserves more attention than hitherto given.

Sources and Structure

Natural philosophers, hydrographers, and navigators were all in search of discernible patterns to make sense of their world, either on a local or a global scale. In trying to uncover the relevant characteristics, this research has concentrated on two main types of historical sources: *ships' logbooks*, and *geomagnetic tracts*. The navigational logbook was kept by a ship's officer in charge of navigation, usually a *master* or a *mate*. It generally adhered to a fixed format, with designated locations on a page to store specific types of relevant information. Individuality was the exception rather than the rule; most navigational logs solely excel in the repetitive rendering of observations and calculated position for each nautical day. Although their authors could read and write, and most had received mathematical training, they did not intend to thrill prospective audiences with captivating stories of life at sea. Navigators were concerned foremost with getting safely and speedily from departure to destination. Logs were therefore kept as a tool, to keep track of the ship's progress, and to record useful data for future reference.

Contrasting with the matter-of-fact style of logbooks are many treatises devoted to the Earth's magnetic field itself. Despite profound differences in penmanship and background, their authors shared the intention to convince the reader that only their particular insights bore the exhilarating hallmark of truth. Several ideas spread through privately printed pamphlets and books, while others received attention in scientific periodicals. Some hopefuls generated hundreds of pages with predicted magnetic values all over the globe, while others sufficed with a few mathematical parameters, from which the local direction of the compass needle could be derived everywhere. Roughly a hundred such postulates from the period 1500–1800 have so far come to light. This publication frenzy was driven in part by the promise of financial rewards, issued to solve the outstanding problem of determining longitude at sea.

These great expectations of theoretical advances serve to balance the more down-to-earth concerns of the navigator, as expressed in the logbooks. While early geomagnetists bickered and argued about the number and situation of the world's magnetic poles, most mariners satisfied themselves with less lofty aims of more immediate practical import. Bridging these two worlds is a gray area of very diverse historical sources. Navigational textbooks, for instance, can elucidate to what extent theoretical considerations trickled down to navigational practice; compass patents and ships' inventories can show when certain

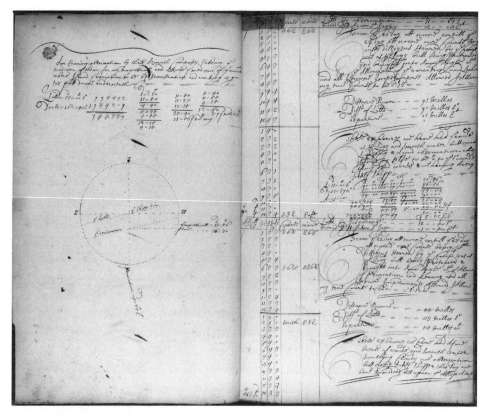

Fig. I.2. Entries for 22 and 23 April 1690 in the navigational logbook of the English East Indiaman *Defence* (IOBL, L/MAR/A 90), sailing from England to Madras. On the left page, a diagram of observed local magnetic declination (4°37′ northeasting). By permission of the British Library, London.

technological improvements came into being and usage; nautical almanacs may hold magnetic data; letters, diaries, and other personal papers can convey opinions held on recently published geomagnetic theories, as well as exchanges or animosity between individual proponents; sailing directions associate values of magnetic declination with specific geographical locations at a certain time. The geomagnetic field indeed pervades many media.

There is, in fact, so much historical source material left in various European repositories that restrictions need to be applied. Clear demarcations in space and time were a prerequisite for this study, not only for successful completion of the research within a certain period, but also to maintain some mental sanity when confronted with a wealth of primary sources. Investigations have therefore primarily been conducted in England, France, and the Netherlands. These three early-modern powers produced a sufficient number of manuscripts to enable comparison between different organizations, epochs, and shipping routes within each individual country, as well as to study the development and spread of geomagnetic thought. Some Danish logbooks have furthermore been added to represent a minor player in the maritime arena. In all, over two thousand processed logbooks have contributed to the findings laid down in this book.

In this analysis, the time range for navigational sources covers the seventeenth and eighteenth centuries. Prior to 1590, the number of extant logbooks describing ocean crossings is simply too low for meaningful analysis and global modeling of the Earth's magnetic field. After 1800, the use of iron and steel in ship construction exacerbated the problem of *deviation*, the deflection of compass needles due to nearby iron. This effect adds a structural and complicated measurement error that is more easily avoided in advance than corrected afterwards, both aboard ship and in interpreting magnetic data. In addition, the many changes affecting maritime organizations around the turn of the nineteenth century (for instance, the demise of the French and the Dutch East India companies, and the fusion of the Dutch admiralties into a single state navy) provided a second reason. A cut-off around 1800 is additionally supported by the geophysical argument that data coverage is both qualitatively and quantitatively better for the nineteenth and twentieth centuries than for previous times. After 1800, a network of land-based magnetic observatories came into being, and eventually spread widely across the globe, while dedicated magnetic surveys with more accurate instruments came to replace datasets compiled by navigators with other priorities than scientific study.

Furthermore, the historical investigation of geomagnetic concepts spans three centuries, from 1500 to 1800. The inclusion of the sixteenth century is necessary because a number of fundamental ideas were first launched at this time in Portugal and Spain, which subsequently found their way northward. Nevertheless, England, France, and the Dutch Republic once again provided the historical skeleton on which most of the conclusions have been fleshed out. In the interest of tracing their spread, geomagnetic hypotheses from other countries than the chosen three were included as well, when and where they were encountered in archives and libraries, even though no active search was undertaken to find them.

Finally, a few words need to be spent on the structure of the following narrative. The presentation of the findings has been split into two major sections, respectively representing the theoretical and the practical aspects of humankind's historical dealings with geomagnetism. Each of these can be read independently, although only their combination offers a comprehensive perspective. The first part pursues the subject within the context of the history of science, identifying four phases of increasing complexity in postulates put forward to explain the observed field. This section is preceded by a brief introduction outlining the present understanding of the geomagnetic field. The second part is firmly rooted in maritime history, and covers the substantial navigational implications of compass use at sea. The final chapter ultimately brings the two parts together, drawing on both disciplines concerned. It shows how a planet-spanning computer model of the field's changes over historical time provides the key to extract new information from old sources, by laying bare quantified differences in the ability to estimate and allow for local magnetic declination at sea.

Needless to say, such an approach requires that focus change constantly, both in time and place. Within much of the first part, a chronological order is adhered to. The chapters on navigation are more thematic in structure, centering as they do on the diverse practices themselves, rather than on their often disparate transformations over time. Furthermore, regarding the countries investigated, national borders appeared ill-equipped to delineate the geographical spread of geomagnetic science and navigation. Regardless of provenance, all seafarers faced the same struggle in coping with the unpredictable behavior of the magnetized needle. Similarly, longitude-finders and natural philosophers kept abreast of the labors of their predecessors and competitors without restricting their survey to their own countries. For these reasons, an international perspective has been adhered to throughout the book. Several trends have parallels on both sides of the Channel, and sometimes direct contacts and exchange have been identified. It seemed more productive to compare their communalities and differences in a single northwest European framework than to generate an artificial, isolated history for each region. In the following discussion, both the Earth's magnetic field and its historical implications for science and seafaring will therefore be considered similar, in that local peculiarities are perceived as idiosyncrasies of dynamics on a far grander scale.

Part I / Earth's Magnetism

CHAPTER 1

The Earth's Magnetic Field

How is one to imagine a magnetic field? It cannot be seen, it cannot be touched, it can only be detected through indirect means. Many people will be familiar with the experiment of iron filings on a sheet of paper, aligned in curves by a magnet underneath. But what of the space between the filings? And how about the region above and below the paper plane? Is the strength everywhere the same? Is it stable or does it change with time? What causes it to take on its peculiar shape? How does one turn an invisible quantity into a physically meaningful picture?

If these questions relate to the Earth's magnetic field, a global answer requires a whole battery of techniques. The discipline of geomagnetism addresses issues ranging from satellite observations to nonlinear equations, and from the magnetization of the sea floor to heat transfer within the planet's core. One of its main aims is to derive the best possible mathematical description of the geomagnetic field over the past few centuries. Assessing the underlying dynamics is vital to our understanding; it is easier to analyze the workings of a clock by watching the moving parts than by studying a photograph. A *time-dependent model* shows the recorded perturbations of the field over historical time, yielding information regarding its origin. These may in turn provide constraints on postulates regarding other processes inside the Earth.

The general composition of our planet has been known for about a century: a stony exterior, formed by crust and mantle, hides a core predominantly made of iron (see fig. 1.1). Separating mantle and core is the *Core-Mantle Boundary* or CMB, at a depth of almost three thousand kilometers. Descending across this profound division, the density almost doubles, as solid rock gives way to molten metal. At the center of this spherical volume of liquid sits the inner core, a gigantic crystallized iron spheroid. The reigning extreme pressures keep it solid despite the high temperatures there. In between this dense kernel and the enveloping mantle lies the source of the Earth's magnetic field: the outer core. This chapter intends to sketch in broad strokes our current understanding of the dynamics at work, enlisting the explanatory prowess of three sea-

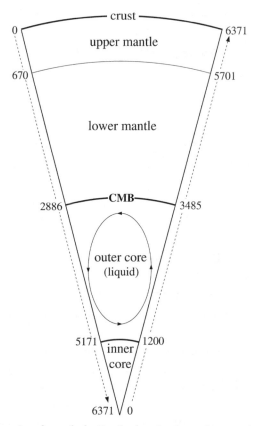

Fig. 1.1. Cross section through the Earth, showing its major constituent parts: crust, mantle, and core, and the depth (left) and radius (right)—in kilometers—of their boundaries.

soned experts: Professor Jack Jacobs (former head of the Department of Geodesy and Geophysics at the University of Cambridge), David Gubbins (Research Professor of Earth Sciences at the University of Leeds), and Kathy Whaler (Professor of Geophysics at the University of Edinburgh).[1]

Core Flow

Understanding the processes that generate and shape the geomagnetic field requires a detailed analysis of the physics at work in the fluid outer core. To start with, one has to imagine the intense heat in the deep Earth, between three and six thousand degrees centigrade, trapped within a thick insulating layer of mantle rock. Our planet is still cooling down from the time of formation, while additional heat is released by the decay of radioactive elements, chemical processes, and liquid material solidifying onto the surface of the inner core.[2] Secondly, in a fluid such as molten iron, thermal buoyancy drives *convection*, transporting the heat to overlying cooler regions. Thirdly, as the world turns around its axis, the whole of this white-hot cauldron is permanently spinning round, forming a rotating compound structure of a solid sphere (the inner core) inside a liquid sphere (the outer core) inside a second solid sphere (the mantle). This system of nested shells already creates quite complex patterns of fluid flow. The physical forces in such a rotating frame of reference then introduce additional twists in the convective motions, deflecting rising fluid to rotate counterclockwise in the northern hemisphere and clockwise in the southern. This process creates slowly spinning rolls parallel to the Earth's axis, and a large ring-shaped region of flow aligned with the equator.[3]

So far, we have a spherical pressure cooker roughly the size of the planet Mars, full of rather hot soup, constantly heated from the center. Now imagine this soup to be extremely well able to conduct electricity. A sufficiently large volume of a conducting fluid, rotating and convecting, is able to form and maintain a *magnetic dynamo*. All it needs to start is the presence of a small magnetic field, for instance, as a relict from the time of the Earth's formation. While heat and rotation propel the molten iron, a potentially small initial magnetic field will generate an electric current in the conducting liquid interior. An intricate pattern of current loops then forms a secondary magnetic field, which will interact with the fluid motion to amplify the original field. The induced electric currents would long ago have dissipated due to electrical resistance if the process did not constantly regenerate itself. The *geodynamo* converts fluid motion into moving electric charge into magnetic force, and will remain self-perpetuating for as long as liquid iron continues to convect around the inner core.[4] David Gubbins explains:

> The device is exactly the same as a car alternator. So you have a magnetic field produced by an electric current in a coil, and then you spin an armature inside that, which induces an electric current, and that electric current is used to charge up the battery in the car. But part of it is also used to go through that coil, to make the magnetic field that is needed for the alternator to work. So it is self-generating in a way; it generates its own electric current . . .
>
> In the Earth's core, everything is the same, except we do not have this complicated mechanism of brushes and coils and so on, we just have fluid flow mov-

ing in a rather complicated manner. The fluid flow is ultimately driven by heat . . . and it is the [Earth's] rotation that provides the essential organizing component that allows the fluid flow to move in the right way to generate the magnetic field.[5]

The science of magnetic fields in fluids is officially called *magneto-hydrody-namics*, and is not for the faint of heart. A realistic causal model would have to take into account electromagnetism, hydrodynamics, thermal aspects, geometry, physical constraints, and all interactions between them. The resulting avalanche of mathematical equations bears some resemblance to a model of long-term weather; both combine many unknowns with constantly changing features and complex instabilities. The meteorological analogy needs to be scaled in time, though, due to the slow velocities of core events compared with those in the atmosphere. Geomagnetic disturbances, due to turbulent surges in the liquid iron, grow over decades and may persist for hundreds of years. One day of "core weather" would last approximately half a century.[6]

The concept of the fluid geodynamo is of relatively recent origin; important contributions date back to the 1950s. Jack Jacobs recalls: "The geodynamo started essentially through the work of Sir Edward Bullard and [Walter] Elsasser . . . Earlier, Sir Joseph Larmor had suggested that the magnetic field of the Sun might have operated on a dynamo, but it was essentially Bullard and Elsasser, in fairly highly mathematical papers, and after that, people almost continually worked on it . . . It gets more and more sophisticated, the more one knows about it."[7] Initially, many obstacles had to be overcome to get to grips with the subject matter. For one, modeling a real magneto-hydrodynamo in a tank (one pursued line of inquiry) is far from easy. Kathy Whaler points out the difficulties of setting up laboratory experiments: "The main problem is trying to find a magnetic fluid that is safe to work with, and not phenomenally expensive. It is hard. One group uses liquid gallium, which is opaque, so it is very difficult to visualize what is going on. The other thing that people are starting to use now is liquid sodium, which is an incredibly dangerous substance to work with. Some of the old laboratory dynamo models involved blocks of metal that rotated, and they would be lubricated by mercury (providing the electrical contact), which again was not a particularly healthy environment to be working in."[8]

Fortunately, numerical simulations on large computers provided a safer alternative for studying both the observed field and fluid dynamo behavior. However, the complexity and sheer scale of the calculations involved placed extreme demands on computing power and system memory at the time. Gubbins paints a picture of available facilities in the early 1970s: "The computer . . . had its own building . . . and I remember it had a memory of . . . 200 kilobytes. Because I had a big calculation to do, I got a five-megabyte disk, which looked like eight 78-records stacked on top of each other, that you screwed into a disk drive about the size of a washing machine. And we used punch cards, and you could get three, five turn-arounds a day. In other words, you could only make five mistakes in the code . . . per day."[9]

Another problem warranting serious attention was data coverage, both in a geographical and a temporal sense. In order to make a reliable map of the global field as it changes over time, a geophysicist would ideally require automatically processed, continuous measurements, recorded all over the world. Jacobs recalls a very different reality: "If you have seen maps of where the magnetic observatories are, they are nearly all in the northern hemisphere, and for that matter, nearly all concentrated in Europe and America, so you have a very uneven net . . . I remember that I used to look at a lot of the data . . . essentially it was all done by hand . . . I used to do some of it myself, with a little magnifying glass, and you put it to your eye, and try to line the thing up. But it was very, very slow, and very, very tedious."[10]

All of this changed with the advent of satellite technology, in particular after the launch of *Magsat* in 1979. Whaler, who was still a Ph.D. student at the time, remembers: "This was a really strong influence on . . . our ability to get good maps of the field at the top of the source region . . . Once we had done that with satellite data, where the coverage was more or less uniform, so that we knew that we were actually able to resolve the features that were appearing in these maps . . . then we were able to trace back in time. We started to see some continuity in the features that we saw in maps based on satellite data and those based on earlier data."[11]

Since then, the several scientific communities that together make up geomagnetism have pulled their resources together, in a concerted attempt to fathom how the geodynamo actually works, using both sophisticated mathematics and observations from the present and the distant past. Whaler asserts: "We are able to simulate the geodynamo in a much more realistic fashion than we were twenty-five years ago. At the time I was doing my thesis, there weren't very good connections between the geomagnetists . . . and the applied mathematicians working on geodynamo theory. A lot of it was simplified analytical models . . . not very well connected to the geomagnetic data . . . Since then, there have been a lot of advances in sorting out what we can actually deduce about the flow of the liquid iron."[12]

The Time-Averaged Field

The simplest magnetic field imaginable is that of a bar magnet, or a uniformly magnetized sphere. It has two poles where the field lines bundle together, and is hence designated a *dipole* field. If the Earth were to have such an ideal dipole aligned with its rotation axis (a so-called axial dipole), then its lines of magnetic force would trace smooth, curved paths, exiting near the geographical south pole and entering near the north pole (or the other way around). An important concept in describing such field patterns is magnetic *flux*. Whaler explains: "An easy way of thinking about it is in terms of field lines, the sort of things that you see if you put iron filings on a piece of paper over a bar magnet and tap it—then you can actually see the field line pattern. The concentration of those field lines measures the strength of the field, and that is what I mean by *flux*: the number of field lines that are enclosed in a little volume, or poking out through the Core-Mantle Boundary. Hence we have *high-flux* regions where lots of field lines are leaving or entering an area."[13] Thus, if the Earth's field was simply dipolar, the whole of the southern hemisphere would contain outward magnetic flux, the northern hemisphere only inward flux. Moreover, the field intensity would be lowest at the equator, and higher toward the poles, where the field lines are close together. The geographical equator would furthermore coincide with the *magnetic equator*. Everywhere on this line, the field exerts a force parallel to the surface only; no flux enters or exits there.

The dipole best fitting the presently observed situation is tilted about eleven and a half degrees away from the geographical poles. The *geomagnetic poles* are located where this tilted dipole axis cuts the Earth's surface, whereas the *geomagnetic equator* is defined as the line where the surface intersects a plane perpendicular to the dipole and through the center. These two concepts need to be distinguished from their counterparts defined by the local field direction: at a *magnetic pole* or *dip pole*, the field direction is solely vertical, while anywhere on the *magnetic equator* it is completely horizontal.[14] Two, and only two, geomagnetic poles exist for a given dipole, but a real magnetic field can have many areas where the field lines run perpendicular to the surface.

Reliable maps of the actual magnetic field at the Core-Mantle Boundary, made by projecting surface measurements down to the top of the core, have been available since the 1980s, and typically show far more complexity than a dipole field would. Gubbins puts it this way: "When you go inside the Earth, down toward the surface of the liquid core, then you see the magnetic field becoming much more complicated because you are closer to the source. Up here, we are a long way from the source, so you just see a north pole and a south pole, but as you go down to the core surface, you actually see the magnetic field concentrated in a lot of regions."[15]

Various descriptive models have since been calculated, displaying a bewildering array of features. Some of these grow, move, and decay again, while others appear virtually unchanging. Interestingly, even the constant parts of the field do not look like that of a simple dipole.[16] Several studies have tried to define this "standing part" of the field through analysis of the time-averaged situation over millions of years, based on magnetized lavas and sediments. When all variations over time are thus ignored, the remaining field at the Core-Mantle Boundary shows four regions of intense force. These do not coincide with the geographical poles, but merely rim the polar regions, near 60° north and south latitude, and 120° east and west longitude respectively. Geomagnetists like to call these two pairs of concentrated field lines *flux lobes*. Remarkably, the geographical poles, where a dipole would generate the highest magnetic intensity, actually exhibit very low field strength. The stability of these features over geological time seems to suggest long-term core-mantle interaction, because the molten iron itself flows far too quickly (at fifteen to thirty km per year) to maintain these patterns over millions of years.[17]

So how could the fluid flow at the top of the outer core be influenced by the overlying solid mantle? Geomagnetists have proposed three possible mechanisms:

1. Inverted "mountains" of rock protruding down from the mantle into the liquid core may affect the flow.
2. Variations in the electrical conductivity of the mantle could locally alter the magnetic field.
3. Differences in heat transfer through the mantle may cause the temperature at the boundary to vary, affecting convection.

Other factors operating on the interface may further complicate the flow in the outer core, such as mantle convection, chemical processes, and partial mantle melt.[18]

Geophysicists postulate that the flux lobes constitute evidence of the geodynamo at work. They interpret the four patches as the tops and bottoms of two columns of liquid, running parallel to the Earth's rotation axis and just touching the inner core. The fluid is thought to spiral down through the two rolls from the Core-Mantle Boundary toward the inner core; this process concentrates the field lines in bundles. In addition to these downward-directed columns, they expect two others along which the flow would be away from the center. Such a roll could be situated midway between the two downward-flowing columns, near 180° longitude (under the Pacific Ocean), where magnetic flux is actually observed to be very low.[19]

The location of the four main lobes seems to suggest a missing third pair of columns near the Greenwich meridian, which would result in threefold symmetry (six rolls: three flowing toward the inner core, and three flowing away from it). However, the missing third pair would reside under the Atlantic Ocean, where active disturbances currently roam (see below). The associated vigorous fluid motions in the outer core may therefore have redistributed the flux, preventing the formation of rolls in this region.[20] It is these disturbances that are responsible for most of the mariners' problems regarding compass use in early-modern times.

Secular Variation

Once the safe footing of a time-averaged view is relinquished in favor of a chronological perspective, the full dynamic picture emerges. If the static field is "subtracted" from the observed magnetic field over time, the remaining variations run the gamut of timescales, from millions of years right down to fractions of a second. Some of these are periodic; others appear to be random. Most long-term effects result from the flow patterns in the outer core. Those changes of internal origin that exist for years and decades to tens of thousands of years are called *secular variation*.[21] The geomagnetic field's continuing alterations provide a convincing reason why none of the solid parts of the Earth (crust, mantle, or inner core) would be able to generate the observed field.

Studies of secular variation are based on measurements made in the last few centuries, and have yielded important clues about the Earth's inner workings. Gubbins mentions some of the findings:

> We can see things in the historical data that are quite surprising: for example, the lack of any change in magnetic field in the Pacific, which seems to be supported

to some extent by the paleomagnetic observations; the rapid changes in the South Atlantic and Indian Ocean region are reminiscent of things that are happening on the Sun, so that gives us an analogy we can work on; and also the fall in strength, which . . . appears to be a small part of a long drop that has been going on since Roman times . . . That is very important, because it indicates that this . . . is a normal phenomenon . . . The magnetic field strength is never stationary; it is continually waxing and waning.[22]

When one compares the field at the Core-Mantle Boundary with that of a standard dipole, various characteristics come to the fore. Apart from the unchanging features already spoken of, the magnetic equator does not coincide with the geographical equator. Moreover, several areas exist where the field strength is either substantially more, or less, than would be expected from a dipole field. Some of these disturbances seem fixed over historical time, such as a permanent spot of very low magnetic force under Easter Island. Other peculiar nondipole aspects shown in the models are apparent field oscillations under Indonesia. A succession of highs and lows under Europe and the North Atlantic Ocean is likewise suggestive of wavelike behavior.[23]

Expanding the temporal window to encompass the last few centuries, and confining attention to the Atlantic hemisphere (between 90° W and 90° E longitude), the phenomenon of *westward drift* is unmistakably apparent. Field anomalies are there observed to be traveling west at roughly a degree of longitude per five years. However, this is not a global phenomenon. These displacements are notably absent from the rest of the world, a fact which invalidates the formerly widespread belief that the whole of the outer core somehow rotated relative to the mantle.[24]

Looking more closely at the drifting disturbances, scientists have found they invariably seem to come in pairs. Their formation may be due to forces associated with the planet's rotation, which can stretch field lines, wrap them around the rotation axis, and cause them to form a large donutlike bundle parallel to the equator. Hot rising fluid in the outer core may then push this ring-shaped field outward beyond the Core-Mantle Boundary, forcing it to locally enter and exit the mantle. This expulsion concentrates the field lines, producing both a high-flux and an intense reverse-flux feature, which may survive for hundreds of years. This combination is magnetically similar to a sunspot, and hence has been baptized a *core spot*. Such a pair of intense flux areas of opposite sign can grow quickly, and may subsequently drift thousands of miles in the fluid interior.

But despite the scope and power of these explanatory visions of the internal Earth, a word of caution seems justified here. One has to clearly distinguish between real, observed features, as captured in empirical, descriptive models or *field maps*, and yet-to-be-proved, postulated causes, as incorporated in computer simulations.[25] Although the concept of the fluid dynamo itself is no longer contested among geomagnetists, large differences in interpretation remain, and vital parameters are either wholly unknown or merely an educated guess. Furthermore, given existing mathematics, the nonlinear nature of the geodynamo precludes our ever being able to confidently predict future field changes. Our understanding of the mechanism becomes even hazier when the behavior of the field over geological time is taken into account. Therefore, much basic geomagnetic research remains to be done, as Gubbins frankly admits:

> The numerical simulations I look at . . . are being run for very unrealistic parameters. In other words, some of the properties that we know are right for liquid iron have been changed, because we cannot possibly solve the equations we have to solve if we use the right numbers. So we changed them to make them simpler, by many factors of ten in some cases, and yet these computer simulations still look to me to be more complicated than the real Earth . . . So I think the real challenge now is to get a model that is as simple as the Earth appears to be . . . We do have another decade or two to go before we are really going to understand the most important phenomena that are going on.[26]

Whaler points the finger in particular at the remaining technological limitations: "It is really a problem of computing power . . . We could put the right parameters in, and I think the programs would probably still work . . . but they would just run so slowly that in any of our lifetimes you would not be making any progress at all in getting useful results out."[27]

Magnetic Measurements

Far from the conundrum of Earth's seething interior, the magnetic field at the surface displays a more gentle side of its disturbed nature. Muffled by the intervening distance of the blanketing mantle, the extreme features of the outer core have their lines softened, their contrasts toned down. Here, neither flux lobes nor drifting core spots litter the magnetic landscape. Instead, by a first and crude approximation, the field looks somewhat like a dipole. This is not to say that the above-mentioned processes in the deep Earth have no relevance to the crust and its inhabitants. Secular variation affects the surface field to a considerable degree, and many sailors may have perished by insufficiently accounting for it. The dipole approximation accounts for about 90 percent of the surface field; secular variation is responsible for most of the remaining anomalies in direction and strength.

In addition, a closer inspection of magnetic readings reveals the influence of other substantial sources of magnetism besides the core. What is gained by attenuation with distance from the source turns out to be partly lost again by new disturbing elements in and above the crust.[28] For instance, all ocean floors demonstrate a tapestry of magnetic anomalies, weaving strands of alternate polarity into an intricate pattern, which may locally affect the compass in shallow waters. In addition, large bodies of iron ore buried in coastal regions may lure the magnetic needle as well. At short distances, even a big rock struck by lightning can cause a significant deflection.

Lastly, several current systems outside the planet are generated by the interaction of the geomagnetic field with the *solar wind*, a variable stream of ionized gas carrying the Sun's magnetic field. Although the resulting magnetic fields form in the atmosphere and beyond, they can under certain conditions affect ground-based observations as well. This happens when the solar wind suddenly intensifies, for instance due to sunspot activity or solar flares. The ensuing particle flow from the Sun is known as a *magnetic storm*. Such an event usually includes the formation in high latitudes of *aurorae* (the northern and southern lights). These harmless, spectacular displays occur when the Earth's field captures some of the Sun's charged particles, and draws them toward the poles. There, they ultimately collide with oxygen and nitrogen atoms in the atmosphere, which causes colorful lights in the sky, forming arcs, curtains, streamers, and diffuse patches.[29]

All aforementioned sources of magnetism may affect a surface reading. In order to separate these various influences, it is necessary to quantify the observations in a mathematical framework. So how do geophysicists actually do this? Picture for a moment the weather forecaster's map, with scattered symbols of Sun and clouds. At several places, the observed or expected temperature is denoted by a certain value. Underlying this representation is a conscious simplification, based on an unspoken understanding between meteorologist and audience. Both know that the absence of readings for places other than the featured localities does not imply that the temperature there is either absent, unknown, or zero. Implicitly, the number of degrees between two known quantities is assumed to be estimable by interpolation. It is a practical scheme, which works because it is based on thermodynamics, prescribing that temperature will change along a gradient, without gaps or randomly fluctuating values. To each point within the region corresponds a single value.

A quantity such as temperature, with magnitude only, is known in mathematics as a *scalar*. Instead of giving a tediously long list of temperatures for each coordinate-pair on the map, it is possible to fill the gaps in between the recorded point values with the aid of mathematical techniques. Temperature is then considered a *scalar function* of position. In other words, it has been defined as a *scalar field*. Other weather-related examples include air pressure and levels of rainfall or smog. All can be approximated by scalar functions, which yield a single quantity for each entered set of coordinates.

A similar equation could represent local wind speed. However, wind force is fundamentally different from the previously listed phenomena in that it has both magnitude and direction. In the televised weather report, animations depict hordes of arrows chasing each other across the skies, growing in number and size as they curl toward the center of a depression. Indeed, wind can be imagined as a *vector*, represented by an arrow of given length (strength) and orientation. Analogous to what has been said above of scalar fields, it is possible to mathematically define a *vector field* by means of a *vector function*, an expression that defines a single vector for each point within a region. In the case of wind, the general public, living at the bottom of an ocean of air, is merely interested in the horizontal direction. Returning to the Earth's magnetic field, an ordinary mariner's compass likewise rotates only in the horizontal plane, due to a specially positioned weight that counterbalances the vertical magnetic force. Nevertheless, one has to keep in mind that the magnetic vector is actually three-dimensional, pointing up or down, as well as somewhere on the horizon.

Such a three-dimensional vector can be represented in a rectangular coordinate system (x, y, z) with its initial point at the origin (see fig. 1.2). The coordinates X, Y, and Z of its terminal point (the "arrow head") are called the *components* of that vector in the x, y, and z direction; they are themselves vectors along these axes. The horizontal components X and Y have been renamed "east" and "north" respectively, while Z represents the radial force relative to the source, deep inside the Earth. Adding these three together yields F, the complete magnetic vector. Its length equals the total magnetic force, or *magnetic intensity*. The easterly and northerly component, moreover, combine into

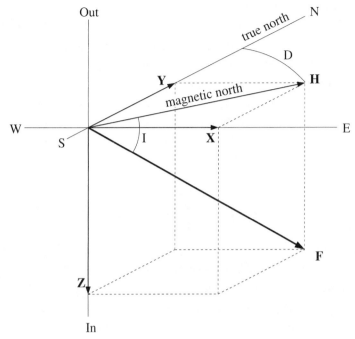

Fig. 1.2. Geomagnetic components in a local reference frame. **X**, **Y**, and **Z** denote eastward, northward, and inward components respectively (negative values designate westward, southward, and outward direction); **H** represents the horizontal, and **F** the full magnetic vector; *(D)*eclination is the horizontal angle between **H** and **Y**; *(I)*nclination is the vertical angle between **H** and **F**.

the horizontal field vector **H**, which describes the local *magnetic meridian,* that is, *magnetic north.* A compass needle will try to align itself with **H**. The horizontal angle it makes with true north is defined as *D,* the *magnetic declination,* which mariners often called *magnetic variation.* The vertical angle that **F** forms with **H** is referred to as *I,* the *magnetic inclination,* or *dip.*[30]

A magnetized needle will try to align itself with the ambient magnetic field. Acting as a weather cock in the wind, it will try to remain tangent to the magnetic flux at all times. And just as the wind does not blow directly from a high-pressure to a low-pressure area, the geomagnetic field does not "flow" directly from pole to pole. A tiny magnet therefore does not point its extremities toward the Earth's distant magnetic poles, but follows the local magnetic direction instead. As intimated earlier, several sources conspire to make this vector point away from geographic north. These include the Sun's magnetic field, the

Earth's external field, and crustal magnetization. The main cause, however, is the internal field itself, due to outer core convection and various types of mantle interaction. The observed local magnetic direction at the surface is therefore usually more easterly or westerly, more outward or inward, and stronger or weaker than would be expected from a purely dipolar field. Thus, a compass that points to true, geographic north is the exception rather than the rule.

In Europe, magnetic declination was probably discovered early in the fifteenth century; awareness of inclination followed about a century later. This vertical measurement required a special instrument called an *inclinometer* or *dip meter*. It looked like a compass on its side, consisting of a needle pivoting in the vertical plane on a horizontal axis. In order to work properly, it had to be aligned with the local declination first, as it measured in the vertical plane through H and Z.

Given the angles D and I, the direction of a magnetic vector is fully defined. Unfortunately, determining its absolute strength (the length of the vector) had to await the invention of the *magnetometer* by C. F. Gauss in 1832. During the two and a half centuries before that time, the horizontal and vertical angles were the only available observations of the local field. Its magnitude remained outside the realm of practical experience, apart from the finding that compasses were generally unreliable in the Arctic. The needle's unsteadiness there results from the fact that a compass is made to be sensitive only to H, the horizontal combination of X and Y, whereas near the poles the field is mainly vertical, so most of the field strength is then seated in the Z component. The weaker the intensity in the plane of measurement, the less accurate a reading will be, and the longer a needle displaced from its preferred orientation will take to return to it. This constituted a severe handicap for early mariners sailing in high latitudes.

Nowadays, historical measurements of the field constitute a veritable treasure trove for geomagnetists, enabling detailed field studies over a time span of about four centuries. Gubbins states:

> The historical observations have turned out to be extremely important . . . because of the way the core moves. The iron . . . cycles all the way around in something like five hundred or a thousand years, and the historical record is four, five hundred years long . . . People often regard the four centuries as roughly equivalent to a week of weather observations, and a week of weather observations is, of course, a lot better than an hour . . . By combining the detailed knowledge of

the last four hundred years with the rough knowledge we have . . . from the previous thousands of years, I think it is going to help us a lot in unraveling this.[31]

Nevertheless, the most powerful computers and the sharpest minds have yet to derive a realistic approximation of the geodynamo, despite ever-increasing datasets of past and current field change. Now imagine a time when most of our present technology and mathematics had yet to be invented; a period when a mere handful of observations took years to reach researchers; an age when next to nothing was known for certain about the interior of the Earth. As a result, both the peculiar distribution of magnetic declination over the globe and its unpredictable changes over time represented huge challenges to navigators, hydrographers, and natural philosophers alike. Their individual and collective responses to a natural phenomenon that refused to be brought into simple rule have been both varied and creative, and hold many important clues about navigational practice, the processing of hydrographical information, and the early development of the geomagnetic discipline. The present insights now offer a unique opportunity to compare the evolution of the field itself with that of theoretical and practical advances made in early-modern times. The rewards of merging history and geomagnetism beckon to be explored, which is the main objective of this book.

CHAPTER 2

The Age of Diversity

Geomagnetism before 1600

Over the centuries, the quest for a rational explanation of the geomagnetic field's observed features has captured the imagination of many. Time and again, navigators, hydrographers, natural philosophers, and laymen have sought to develop theories that would adequately account for an ever-growing body of magnetic measurements. Some of these concepts were purely descriptive, while others addressed underlying causes as well. The field was thought to originate in the heavens, on magnetic islands, in the Earth's continental crust, or deep inside its interior. The complexity of explanations tends to increase with time, due to larger available datasets and more sophisticated methods of analysis. After the discovery of time-dependent change, for instance, primitive, static notions began to give way to a bewildering array of magnetic poles revolving around the geographical axis over centuries. The compilation of datasets of declination covering large areas of the globe similarly opened minds to the possibility of more complex arrangements of multiple poles, at different coordinates and depths, and no longer necessarily paired in diametrical opposition.

A substantial amount of effort and resources was dedicated to the problem: ships were sent out on fact-finding missions, hundreds of logbooks were painstakingly examined for clues, and readings were taken on land under controlled conditions at regular intervals. Natural philosophers corresponded and debated on the findings, while "unrecognized geniuses" spent their often meager earnings on magnets, instruments, and the publication of predicted values tabulated in backrooms. Large tomes resulting from decades of scholarly investigation now stand side by side on library shelves with flimsy pamphlets announcing the definitive answer. All bear witness to the perseverance of this motley crew of geomagnetists *avant la lettre*. None was even remotely right.

The three main reasons for all this activity were the improvement of practical navigation, scientific curiosity, and fortune and glory. The mariner's never-ending struggle to come to terms with the vagaries of the compass needle will be treated at length in the second part of this book. Although local declination could to some extent serve seafarers as a positional indicator, a constant vigil had to be kept both during a single voyage and over the decades spanning a seafaring career. Wherever local declination was observed or known to be other than zero, a commensurate compensation had to be applied in steering, dead-reckoning, and charting. Because the spatial and temporal changeability of the field could not be adequately forecast, navigators and hydrographers always remained at a disadvantage, constantly having to catch up with current events. The merits of a comprehensive theoretical framework were therefore evident, especially in the age of sail, when no alternative to the magnetic compass was available.

Scientific curiosity likewise constituted a legitimate and important rationale for geomagnetic investigations. Apart from being one of the major unsolved problems of natural philosophy in its own right, it had a bearing on (then) related disciplines such as magnetism, electricity, geology, astronomy, and cosmology. Characteristics of the Earth's magnetic field have been employed to defend and oppose several schools of thought (to be discussed), ranging from geo- and heliocentrism, tiny particles, and theories featuring all-permeating emanations. There clearly was much to be gained by a better understanding of the principles and forces at work.

The third reason, fortune and glory, was directly linked to the problem of determining longitude at sea. Ocean travel involves navigation off the continental shelves, without the benefit of landmarks and *soundings* (that is, depth measurements) to establish position. And although latitude could be determined fairly accurately using astrolabe, quadrant, or cross-staff, measuring a vessel's longitude remained an insoluble practical problem up to the 1760s. An early-eighteenth-century manuscript entitled *Sayling by the True Sea Chart* explains why: "That dimension of the Earth . . . which we have termed its longitude or length hath neither beginning nor ending established in nature . . . [T]he first meridian . . . may be indifferently placed anywhere, ther being nothing in nature on the Earth consider'd with relation it might have with the heavens, which obliges us to place it in one place rather than in another."[1]

Longitude has to be measured from a fixed reference. It is different from latitude in that any pair of opposite meridians forms a (vertical) *great circle* (a line dividing the Earth into two equal halves), whereas the equator is the only great circle cutting the globe horizontally. This unique dividing line was thus an obvious choice as global reference for latitude. However, no such agreement could be found regarding longitude. There was no physical or astronomical reason to prefer one prime meridian over another, and no fixed reference point existed in the revolving skies to measure longitude from. The Sun, the Moon, the planets, the distant stars, all pursued their lonely paths across the heavens, oblivious to the mariner's plight. Nevertheless, astronomers put forward several different solutions as far back as the sixteenth century. Sadly, all of these worked only in principle; instruments and astronomy proved to be still wanting in the required accuracy for practical implementation in the sixteenth and seventeenth centuries. In the present context of the development and spread of geomagnetic hypotheses, it is important to keep in mind that finding the longitude at sea remained a practical and acute problem from the beginning of ocean faring until the late eighteenth century.

As a solution could have significant strategical, territorial, and commercial implications for whichever seagoing nation would find it first, it is not surprising that several maritime powers actively promoted the search for this navigational Holy Grail. As early as 1567, Spain's King Philip II issued a sizeable reward, reiterated by his successor Philip III thirty-one years later. In the Northern Netherlands, both the Estates General and the Estates of Holland promised monetary encouragement, consisting of five thousand guilders by the former authority in 1600, and three thousand pounds reward in 1601 by the latter. About a decade later, the Estates General even increased their offer to fifteen thousand guilders. Furthermore, in 1714 the English Board of Longitude came into being by official act.[2] Consisting of twenty-two naval officers, scholars, and statesmen, this committee was to assess all proposals pertaining to the determination of longitude at sea, as well as other submitted improvements in practical navigation. It could award modest prizes for minor suggestions, and larger sums for substantial contributions. The jackpot consisted of twenty thousand pounds sterling, a staggering amount of money at the time. It was to be issued to the first inventor of a proven reliable method that was accurate to within thirty nautical miles. Smaller awards were promised for accuracy to within forty and sixty nautical miles. In France, no comparable, state-funded equivalent to the board existed up to 1795, although the Paris Académie des Sciences did examine several potential solutions. Furthermore, both in England and France, some private individuals bequeathed funds to endorse research in this area, providing additional stimulus.[3]

These substantial rewards doubtless set many an eye atwinkle. Although the Dutch prizes elicited only ten serious proposals during the first half of the seventeenth century, the English board could boast over two hundred supplicants at the end of its tenure in 1828. Unfortunately, quite a few of the submitted schemes were either impractical, based on erroneous assumptions, or simply ludicrous. The lure of financial gain seems at times to have had a debilitating effect on the self-criticality of some applicants.[4]

In the countries surveyed in this study, about one hundred of the proposals scrutinized by experts or simply unleashed into the public domain concerned geomagnetism. The majority directly linked postulated field regularity with navigational application (to determine either longitude, latitude, or both), whereas a minority followed a more theoretical skein, sketching a descriptive or causal model. Of the many that have been laid before official bodies for appraisal by experts, only a few have actually been tested at sea, after temporarily being given the benefit of the doubt. But none was ever accepted by the majority of navigational practitioners as a permanent solution to the longitude problem.

Regrettably, several proponents merely alluded to their schemes in general, without rendering any particulars other than that the Earth's magnetic field was involved. Some apparently feared the jealousy of their competitors; in other cases, specific details may have been communicated but subsequently lost. Whatever the reason, the historian of science is left with a number of enticing promises without enough substance to define their characteristics. Consider, for example, ship owner James Moore, who submitted a hopelessly incomplete geomagnetic longitude proposal to the Board of Longitude in 1790, with these words: "I have constructed an instrum.t which I have great expectation will be of infinite service to navigation. It is so contrived as at one view to shew the variation of the compass, and at the same instant point out both latt.de and long.de."[5] His confidence in his invention was such that, prior to full presentation to the board, he resolved to prove its utility at sea in his brig, the *Maria of Cork*, which he intended to take via Bordeaux and Madeira to the West Indies. As no word was received from him ever since, his laudable but somewhat naïve attempt to test his hypothesis may have backfired in a most dreadful manner.

The Tools of the Trade

"Many learned men write that Nature is hiding four things," states a Dutch navigation tract from 1659, namely, "the movement of the heavens, the movement of the Earth, the movements of the heart, and the movement of the magnet."[6] Indeed, the phenomenon of magnetism was poorly understood at the time. Even though a magnet's ability to attract, repulse, and magnetize had long since been acknowledged, the quantification of magnetic force as a function of distance proved to be a difficult problem. This was not surprising, since the necessary mathematical and physical concepts were still in their infancy. Moreover, a seventeenth- or eighteenth-century natural philosopher was totally unable to measure absolute magnetic intensity. He was able to determine magnetic directions, but not magnitudes; he was living in a world of angles (declination and inclination) rather than vectors. Worst of all, he was unable to mathematically describe a magnetic field in terms of a vector field (which associates a vector with each point in a space). Instead, he assumed a direct influence of the Earth's magnetic poles on the compass needle. This would be equivalent to a cartographer delineating the course of a wildly meandering river by merely indicating its source and mouth. With only beginning and end point to contend with, the river's capricious curves through the landscape could then be reduced to a straight line on a plane chart. Now suppose on a very foggy day, a holiday maker decides to go fly-fishing in that river, armed only with that map. When asked, while standing knee-deep in the water, to indicate the direction of the river's source (or mouth), and with only the flow in his immediate vicinity to guide him, chances are fairly slim that he would be right, and even then only by accident. Around the next bend, away from that big rock, or even just standing nearer to the other bank, his answer could have been different.

Similarly, a compass is guided by the local field characteristics, as generated by a churning mass of liquid iron in the outer core. Furthermore, the Earth's magnetic field does not resemble that of a simple bar magnet of planetary proportions, even though it does have a strong dipolar component. It will be recalled that the two geomagnetic poles are defined as the intersections with the Earth's surface of an axis coinciding with this dipole. However, they are only imaginary points in a mathematical description, and certainly not the hallowed place that all magnetized needles everywhere would respect. This misconception forms the basis of most early modern geomagnetic hypotheses, and lies at the root of their failure to adequately describe, explain, and predict measurements of D and I. The fundamental problem lay in the fact that most concepts assumed an extremely limited number of poles to constitute the sole origins of magnetic force. Thus, the focus of analysis was diverted from the field as a whole to establishing the exact location of these faraway, "all-powerful" points of attraction.

At its initial stage, a popular cosmological interpretation equated the magnetic poles with the *celestial poles,* and thereby aligned the dipole with the Earth's rotation axis. Alternatively, a giant piece of magnetized iron (a magnetic mountain or island) located somewhere in the Arctic was imagined to exert influence the world over. An additional source of information was obtained by cutting and polishing a piece of magnetized iron ore—a *lodestone* (or *loadstone*)—to represent a sphere. Behavior of compasses near such a "little Earth" (*terrella*) seemed to vindicate the dipolar interpretation. Although the difference in size with the real Earth was huge, and the source of magnetism quite different, the analogy was believed by some to hold true.

Imposed regularity received a first blow with the acceptance of magnetic declination as a natural phenomenon. It was thought that local declination had a direct bearing on the position of the dipole, which apparently stood at some angle to the Earth's spin axis. On a sphere, the shortest line between two points is part of a *great circle* (the intersection of the surface with a plane through the Earth's center and both points, dividing the globe in two equal halves). By applying spherical trigonometry, two observations of magnetic declination at places far apart in longitude could provide a cross-bearing of two such great circles, each through one of the points and at the given angle to true north. Where the two lines crossed, the geomagnetic pole was supposed to lie. Once the position of the dipole was known, a similar calculation would yield the longitude of any place where declination and latitude had been measured. Thus the longitude problem would be solved.

But before delving headlong into the arena of claims, counterclaims, muddled arguments, and obscure dog Latin, it seems useful to briefly discuss the rules of the game. Certain terms and concepts will appear frequently hereafter, and deserve explanation free from the clutter of particular interpretations. Recalling the previous chapter, a dipole coincident with the rotation axis is known as an *axial dipole,* whereas a dipole standing at an angle to this axis is denoted a *tilted dipole.* The extent of this tilt is expressed by the dipole's *colatitude,* that is, the distance in latitude, measured southward from the geographical north pole (at 0°) toward the equator (at 90°) and the south pole (at 180°). Its orientation is fully defined once its longitude is also specified, determining

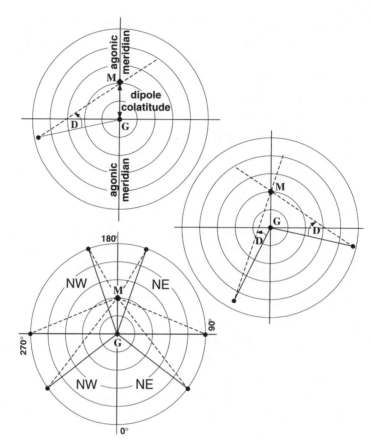

Fig. 2.1. A tilted dipole in polar stereographic (equal-angle) projection, showing *(G)*eographic pole, *(M)*agnetic pole, and *(D)*eclination; the dipole's colatitude is measured from *G* to *M*. *Top left*: on the dipole's longitude and on the antimeridian (at 180° E), a compass would point true north. *Bottom left*: elsewhere, *D* would vary with latitude and longitude. *Middle right*: a cross-bearing using two declinations observed at widely distant locations, intersecting in *M*.

the plane of tilt. Unfortunately, a wide variety of *prime meridians* (from which longitude was measured) were used in the past, so most concepts require conversion to the current, Greenwich-based system to be compared.

Given a tilted dipole at a fixed position, a surveyor traveling around the globe at the same latitude would see his compass needle diverge from true north as a function of longitude. At the meridian of the dipole's longitude and at the other side of the planet (on the *antimeridian*, 180 degrees of longitude distant), a great circle through the place of the observer and the magnetic pole would cut the geographical pole as well, resulting in zero degrees declination. The largest declination would be reached halfway between those two opposite meridians, near 90° longitude east and west from either.[7]

An easy way to visualize the global distribution of magnetic declination is by drawing lines on a chart through all points with equal declination. Such lines are called *isogonic lines,* or *isogones;* the isogonic of zero degrees is also known as an *agonic line.* A tilted dipole system has only a single agonic, namely the above-mentioned great circle through the geographical and magnetic poles. The line is by definition geographically oriented north-south; it is a *meridional agonic,* or *agonic meridian* (see fig. 2.1). Actual measurements of zero declination near the Azores in the sixteenth century led to the mistaken belief that such an agonic was found, forming a "natural" indicator of a pre-ferred meridian to reckon longitude from. The combination of reigning *north-easting* in western Europe and *northwesting* on the American east coast fur-thermore led to the conclusion that the dipole was situated on longitude 180° E (relative to the agonic prime meridian over the Azores) rather than on 0° (which would have resulted in northwesting in Europe and northeasting in North America). When isogonics were mapped on a chart, an alternative method of estimating longitude could in theory be realized by finding the in-tersection of an observed parallel of latitude with the isogonic of the measured declination, providing that the isogonic ran at a sufficiently large angle to the parallel. Yet another way would be to construct a mesh of isogonics and *iso-clinics* (lines of equal inclination) to find an intersection based on the com-bined observed values of *D* and *I*.

Naturally, the more regular the declination pattern, the better. To compli-cate matters, the Earth's rotation axis is itself tilted about 23°27′ from the per-pendicular to the plane of the ecliptic, in which the planet orbits the Sun. This orientation causes the seasons and the Sun's apparent migration between the tropics of Cancer and Capricorn. Relative to this annual movement, another pair of "fixed" points in the sky is defined as the *poles of the ecliptic,* or zodiac poles. Projected onto the Earth's surface during a daily rotation, they would trace the Arctic and Antarctic circles. Quite a few early longitude schemes chose their dipole's colatitude to lie either directly on, or very near to, these imaginary small circles.

A postulate containing more than one dipole is called a *multipole system.* This can be a quadrupole (four poles), a sextupole (six poles), and so on. Mul-

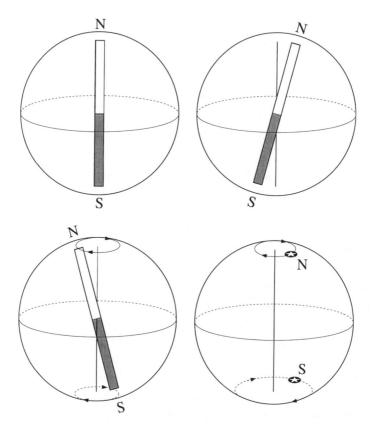

Fig. 2.2. The four phases of geomagnetic hypotheses (up to 1800). *Top left*, axial dipole (static); *top right*, tilted dipole (static); *bottom left*, precessing dipole (dynamic); *bottom right*, disjointed dipole (dynamic).

tipoles are by definition tilted, because if both dipoles were aligned with the Earth's rotation axis, the magnetic characteristics would be indistinguishable from that of an ordinary axial dipole. Unless specifically stated otherwise, each multipole is assumed to be a combination of dipoles, and each dipole to consist of two magnetic poles in diametrical opposition. This means that if the northern part of a dipole is found at latitude x north and longitude y, then its southern end is located at latitude x south, and on longitude y + 180 degrees. Such dipoles are said to be *antipodal*.

A second undermining of simple field representations took place over time, as secular variation manifested itself in observational records, not just as change over time (secular variation), but also as different rates of change at dif-

Table 2.1 The Four Defined Phases of Complexity in Geomagnetic Hypotheses

Phase	Tilted	Dynamic	Antipodal	Dipole(s)
1	No	No	Yes	Axial
2	Yes	No	Yes	Tilted
3	Yes	Yes	Yes	Precessing
4	Yes	Yes	No	Disjointed

ferent times and places (secular acceleration). A new class of dynamic concepts became necessary to account for this newfound temporal dimension, and largely replaced earlier static ideas. The solution most often practiced was to give the dipole(s) a *precession* around the Earth's spin axis. A magnetized nucleus inside the crust, revolving at a slightly faster or slower pace than this enveloping shell, could produce a very gradual displacement east or west relative to the crust, with a period of centuries. In the multipole case, one or several hollow spheres around a kernel, each with its own set of magnetic poles, could be imagined to rotate inside one another at different speeds. The isogonics resulting from such a complex interaction of forces would obviously no longer be meridional, but curved instead.

The imagined path of a *precessing dipole,* as it completes an orbit around the geographical pole, can be plotted. This procedure fixes the direction of the orbit (clockwise or counterclockwise), and its period (usually measured in years). A precessing dipole (which is by definition tilted) is defined by its orbital period and direction, its colatitude, and its longitude at a certain time. The radius of a circular orbit equals the dipole's colatitude.

Lastly, the assumption of *antipodality* was given up in some schemes. This implied that a pole in the southern hemisphere no longer necessarily resided at the same latitude south and 180 degrees east of the corresponding pole in the northern half. Some hypotheses even gave each magnetic pole a completely different rate and direction of rotation around the geographical axis. Several authors argued that a given irregular pattern of declination could equally well be explained by a single *disjointed dipole* as by a multipole of two or more antipodal dipoles.

The above development can be discretized into four phases of increasing complexity in geomagnetic field representation (see fig. 2.2 and table 2.1). These phases represent conceptual, rather than chronological, divisions. Each step constituted the abandonment of a previously adhered to simplification, and an increase in the number of parameters to be determined to match the observations. It does not follow, however, that each stage represents a single period with no overlap, but only that the first attributed instance of each phase preceded those with a higher ranking. Static schemes thus continued to be postulated by some, after the introduction of time-dependent reconstructions by others, and the first rejection of antipodality did not spell the end of all postulated dipoles in diametrical opposition.

All these geomagnetic ideas attempted to recognize pattern in one of Earth's least understood properties. Limitations in available mathematics and data, as well as a lack of knowledge and quantification of the principles involved, confined their proponents to a certain range of solutions, grouped following a few basic tenets, such as the orientation and movement of the dipole. Like most disciplines, the majority of workers associated with geomagnetic conjectures contented themselves with altering existing parameters to suit their particular interpretations, rather than introducing completely novel ones. Arguably, one could write a history of ideas based solely on the few exceptional characters who broke the existing mold. Interesting though these innovators may be in their own right, the purpose of this study is not to present a sequence of "great men" enlightening the world with their genius. In addition, the mere fact that someone was first in committing a new insight to paper does not necessarily make this person the best representative of it.

The four phases outlined above constitute conceptual leaps, which have manifested themselves in various forms. Their intellectual parents can collectively be thought of as a medium supporting the inception, distribution, and competition of postulates. Borrowing an analogy from radio, attention will here be focused on the broadcasts rather than the carrier wave. This approach is not meant to reduce their promoters to hapless participants, swept along by the sheer force of a new explanatory scheme. Individuals will figure where and when their actions directly affect the domain of geomagnetic thought. But stress will be placed on their legacy rather than their lives, on strands of thought —in particular where, when, how, and under what influences they originated, were passed on, mutated, and changed their intellectual environment, and what factors eventually conspired to halt their dissemination. The following history intends to trace the dynamics shaping this peculiar and somewhat neglected intellectual landscape, as it rippled and folded through time.

Early Tenets

The discovery of magnetic attraction in all probability long preceded any written recording of it. Knowledge of the magnet's motive effect on iron may date back to prehistoric experiences of iron smelting. In classical Greece, the idea of two different kinds of lodestone attracting each other seems to have emerged before the notion of the same effect in a single variety of rock took hold. Various classifications attempted to distinguish lodestones on the basis of color, density, and strength of magnetization. The acknowledgment of magnetic attraction was eventually followed by that of repulsion and the transfer of magnetism to iron (the communication of attraction by direct contact of lodestone and iron). No classical sources, however, make mention of either polarity (having poles of opposite sign), or orientation (the tendency of magnetized iron needles to point northward). The concept of a magnetic field was equally unknown.[8]

Left with a range of peculiar experimental evidence, Greek philosophers developed two major, mutually exclusive theories to explain the observed behavior of magnets. The physical one assumed some form of invisible emanation, an *effluvium*, to exude from the "pores" of the lodestone, interacting through mechanical means. By contrast, the competing, metaphysical doctrine worked from the principle of *sympathy*, borne out of a supposed *similitude* between entities. Both will figure briefly here.

According to later authors, Empedocles of Acragas (ca. 490–430 B.C.) constructed a general theory of change based on elemental effluvia. All objects he deemed to be porous, and capable of both emitting and receiving emanations. It was thought that, although air would usually block the pores, iron represented a special case: outgoing flow from a nearby lodestone was assumedly able to dislodge the air particles. As a result, the iron could in turn release its own effluvia, which happened to have a predilection for the pores of the lodestone. The iron then physically followed its own emanations, and was thus seen to be "drawn" toward the magnet.[9]

This seminal construct developed further in the following centuries along two separate tracks. Plato (428–347 B.C.) and Plutarch (A.D. 46–119) took up the original effluvial theory; Plutarch especially emphasized the circulatory nature of the emissions, exiting at one end and reentering at the other. Meanwhile, Democritus of Abdera (ca. 460–370 B.C.) gave Empedocles's mechanistic interpretation more substance. Being an advocate of *atomism*, Democritus replaced the rather hazy effluvia by solid atoms, expelled from the surface of all bodies. The lodestone matter being more "subtle" than that of iron, its emanations were thought to penetrate the iron's interatomic spaces. Forced out, the straying iron particles were then absorbed by the lodestone, and once again the iron was obliged to follow. About a century later, Theophrastus (372–287 B.C.) confirmed the nature of matter to be a "together-flowing" of terrestrial particles, able to act and be acted upon by other bodies.[10]

Near the beginning of the third century B.C., Epicurus (341–270 B.C.) added the notion that the emanations of iron and lodestone tended to unite in a single circulation through both, due to their similarity in particle shape. Another two hundred years later, Roman author Titus Lucretius Carus (ca. 95–51 B.C.) reintroduced the ambient air as an important factor, the circulation supposedly creating a tiny vacuum between the iron and the stone, which drew them together.[11] But despite these differences, effluvial and atomistic mechanisms constituted two sides of the same coin. Both were to reappear in France during the seventeenth century, applied to the Earth and championed by René Descartes and Pierre Gassendi respectively, of whom more will be said later. For future reference, it is important to remember that the term *corpuscular* applies to any theory that relies on the interaction of tiny particles (be it atoms or other *corpuscles*), whereas the physical manifestation of magnetism (what we now call a magnetic field) has been perceived by some as an almost liquid outpouring (*effluvium, effluvial*), by others as more like a gaseous emanation (*ether*). These definitions were neither rigid nor mutually exclusive, and postulated eddies in the magnetic medium further complicated the picture.

The metaphysical alternative to these whirling motions lay in a supposed mutual affinity or dependence of bodies, encapsulated in the concepts of sympathy and antipathy. Early references to these concepts figure, for example, in the works of Pliny the Elder (ca. A.D. 23–79). Slightly later, Galen of Pergamom (A.D. 129–199), who strongly opposed atomism, interpreted magnetism as a "living" force of cohesion and motion. Aristotle (384–322 B.C.) moreover mentioned the opinion of the astronomer Thales of Milete (624–546 B.C.), who allegedly even attributed a soul to the magnet, on account of its ability to move iron.[12] After the fall of the Roman Empire, it was predominantly texts by Pliny, Galen, and Aristotle that remained in circulation. Furthermore, the Platonic doctrine of scientific development through deduction, the placement of observation and speculation on an equal footing, and the reliance on secondary sources compiled by "authorities" resulted in mysticism and explanations based on supernatural *final causes*. These were based on the belief that everything, animate and inanimate, was purposive, being endowed with an internal motive force striving to reach a certain state of perfection. In the case of lodestone and iron, this state constituted a permanent union.[13]

Throughout the Middle Ages, progress in magnetic thought was limited, as Aristotle's commentaries were avidly read, translated, copied, and reinterpreted to fit in with the locally dominant belief systems. In the Christian world, Scholasticism proved particularly creative in this respect, by interpreting magnetization as an example of divine control, or permeation by the Holy Spirit. Furthermore, qualification ranked far above quantification, and, until the late Middle Ages, many medieval authors lacked a beneficial environment for the communication of ideas and data. Apart from the Greek sources, everyone had to practically start from scratch when composing a treatise on natural phenomena. These circumstances maintained the lack of progress, and impeded the swift extrapolation of magnetic concepts to include the Earth itself.[14] Aristotelians assumed that the diffused magnetic "quality" would actuate the iron's self-motion toward the lodestone, and that was the end of the matter. This natural, occult affection was itself not directly detectable, and therefore eluded further analysis, or so the reasoning went. Magnetism furthermore differed from gravity in operating only over short distances, and acting on bodies rather than extending its influence to any point in space.[15]

The doctrine of similitude and dissimilitude as defined by clergyman-scholar Alexander Neckam, in his *De Naturis Rerum*, is basically a late-twelfth-century restatement of these classical maxims of sympathy. Its main innovation lay in the incorporation of the principles of polarity (opposites attract) and orientation. The latter was actually reduced to a manifestation of the former—an incorrect postulate, since a floating magnetized needle aligns with the magnetic meridian, but is not physically attracted toward the nearest geomagnetic pole. This is because the field exerts a couple (two parallel forces of equal magnitude and opposite direction), rather than a single force pulling it in one direction.[16] Despite this oversight, the application of sympathy and similitude to the interpretation of north-seeking behavior finally paved the way for linking the magnetized needle with its environment at large, encompassing the globe or even the whole Aristotelian cosmos. This opened up a whole new field of discussion as to where the invisible pole was located that could impose its will on compasses all over the planet. It thus heralded the beginning of the first phase of geomagnetic thought.

Pole Position

In 1269 the armies of Charles of Anjou besieged the Italian city of Lucera. Among the soldiers was engineer Pierre de Maricourt, better known by his Latinized pen name Petrus Peregrinus. On 8 August of that year he sent a letter to a friend containing an epistle on the magnet, which is a veritable landmark in the history of magnetism.[17] Part of the text concerns the description of two magnetic instruments. The first consisted of an oval lodestone in a vessel of water, kept afloat by two wooden capsules. The second device was the first known dry-pivoted bearing compass. According to the text, its needle oriented itself toward the celestial pole, or true, geographic north. The author explicitly stressed navigational application, either on land or at sea.[18]

Peregrinus also described experiments with a spherical lodestone, on which he had traced great circles. He adhered to the Aristotelian notion of causation, which considered magnetic movement to be a natural quality. He referred to the small sphere as a "round magnet" rather than as a *terrella* ("little Earth"), and with good reason, as he understood it to represent the firmament. Every point on the magnet he considered to have its equivalent in the heavens, the magnetic poles corresponding to the celestial poles. The resulting field would be that of an axial dipole (that is, aligned with the Earth's rotation axis). Magnetic and true meridian will then always coincide, inclination is simply a function of geographical latitude, and declination is zero everywhere.

The sympathetic association between the lodestone and the revolving sky overhead led Peregrinus to postulate a direct dynamic effect as well. While addressing the question from whence the magnetic mineral received its force, the Frenchman described a peculiar experimental setup. It consisted of a spherical lodestone suspended in a frame so as to be able to turn freely about its magnetic axis, oriented parallel to the "celestial rotation axis." Magnetic sympathy between stone and sky he then supposed to force the magnet to follow the celestial sphere in its daily revolution around the Earth. Like its larger mother, the suspended spherical lodestone would turn about its axis in exactly twenty-four hours, like an infallible clock. The idea was to be taken up over three hundred years later by William Gilbert, and subsequently became a bone of contention in the debate between geo- and heliocentrists. These developments will be treated later in this text.[19]

The concept of the axial dipole was reiterated by several authors in the sixteenth century. Circumnavigator Pigafetta, to name but one, in his 1525 treatise on navigation, gave the following rationale for choosing the celestial pole: "The reason of this tendency is because the loadstone does not find in the heavens any other spot in repose except the pole, and on that account directs itself toward it."[20] The Italian moreover presented this intelligence as holding the key to the longitude problem, which is where his explanation became contradictory. A few lines down from his affirmation of the magnet's loyalty to the heavens, he claimed that the measured declination equaled the number of longitudinal degrees between the observer and the nearest agonic meridian. The existence of such a meridian (and of non-zero magnetic declination elsewhere) would actually require the dipole to be tilted away from the Earth's spin axis.[21]

An alternative to the celestial pole itself was the star nearest to it, the Polestar (*Polaris*), in the tail of the constellation of Little Bear (*Ursa Minor*). The same concept of sympathy here established an enduring connection between the star and the compass. This notion perhaps also supported the adoption of the star-shaped rose of the winds on the compass card. Among the Polestar's protagonists were Jacques de Vitry (1218), Thomas of Cantimpré (ca. 1240), Dante Alighieri (1321), and Girolamo Cardano (1550).[22] Depending on whether the Polestar was deemed to be situated exactly at the celestial pole, or a few degrees of arc away from it, the resulting dipole was either axial (phase one) or tilted (phase two). Cardano opted for the second option, estimating the distance to amount to five degrees easterly. This value he could have obtained from an inaccurate measurement of the actual distance at the time (circa $3\frac{1}{2}$ degrees of arc), or an observation of local northeasting somewhere on Earth.

A host of sixteenth-century authors vented less precise ideas concerning the actual placing of the magnetic pole overhead. Most remarkable of these sky-borne tilted dipoles was the one posited by an author of a well-known navigation manual in 1545. Like many others, Martín Cortés remained unclear as to the magnetic pole's exact position. But, given the assumption that magnetic declination did exist (rather than being an error of compass or measurement, as some of his contemporaries maintained), the Spaniard had reached a peculiar, albeit logical conclusion, based upon the reigning cosmological interpretation at the time. This classical philosophy worked from the assumption of a fixed Earth at the center of the stellar universe, inside nine spherical crystalline shells of increasing dimensions. From the inside out, these were supposed to carry the Moon, Mercury, Venus, the Sun, Mars, Jupiter, Saturn, and the stars. The last rotating sphere was known as the "Prime Mover" (*Primum Mobile*), and was supposed to keep the whole system turning by transferring angular momentum to the inner spheres. Given this interpretation of the universe, Cortés reasoned:

— as the magnetic pole is not aligned with the celestial pole, and
— the ninth sphere is constantly revolving,
— the unmoving origin of the magnetic force has therefore to lie *outside* the outermost heaven, in order to remain steady.

The former navigator had thus effectively placed the dipole on the level of the gods.

The Magnetic Mountain

In a copy of the Peregrinus epistle, next to the assertion that the needle respects the celestial pole, a note has been scribbled in the margin. It was probably made by English mystic and mathematician John Dee, in the second half of the sixteenth century, and flatly stated that the French engineer had been wrong on this account.[23] The remark reflected a new insight that was then gaining acceptance among scholars and navigators. No longer was the seat of magnetic power to be conferred to the heavens. Instead, a solitary rock, island, mountain, or a cluster of them, made of the purest lodestone, was supposedly lord and master of all compasses. Once more, this general premise engendered many interpretations. And once again, the roots of the legend go back to classical times.

Claudius Ptolemy's description of Asia (second century A.D.) spoke of the "Manioles," now tentatively identified as near Borneo. The isles were said to exert such a strong magnetic attraction that ships built with iron nails sailing close by would get caught in their pull and be held forever. The fact that local indigenous peoples constructed wooden vessels with fibers added credibility to the story, as their iron-free boats would not be susceptible to the danger. A more colorful version, brewed in Arabian folklore over centuries, had a lone magnetic mountain rising up from the sea that not only lured passing ships to their doom, but physically pulled out all iron nails at a distance, breaking up the integrity of the hull and causing all hands to perish. This powerful image appears for instance in the famous compilation known as the *Book of the Thousand Nights and a Night*, and was similarly conveyed by some Arabian geographers. Agreement on the mountain's location was far from unanimous, but centered on the Red Sea, the Persian Gulf, and the Arabian part of the Indian Ocean.

By the twelfth century, the idea had reached northwest Europe and became incorporated in various epic sagas and legendary voyages, involving Ernst von Schwaben, St. Brandan, Ogier the Dane, Magnus Magnussen, Gudrun, and John Maundevile. Some tales varied the theme by introducing a lodestone castle on an island, or a magnetic rock on the seafloor detaining the ships. Most agreed to disagree on the location of this dreaded peril, although northern waters were favorite.[24]

Two thirteenth-century Italian authors brought the legend one step closer to geomagnetism. Poet Guido Guinicelli of Bologna, in his *Madonna, il fino amore...*, pictured magnetic mountains north of the Alps and underneath the Polestar, which was sending forth invisible force "to turn the quivering needle to the Bear." Elsewhere, he compared love from a woman's eye to "virtue from the star." A contemporary from Padua, Pietro d'Abano, placed the point of attraction further north than the European mainland, which would imply somewhere in the Arctic. It will be recalled that notions of the Polestar as being magnetic began to emerge around the same time. The Italian texts thus presented the legendary mountain as an earthbound receptacle and relay station of this star's magnetic force, distributing it across the globe from the top of the world. It is unclear whether sympathy linked lodestone and magnetic star, or that the stellar influence was supposed to have initially magnetized the rock located directly below.[25]

Furthermore, an apocryphal account of Oxford friar and mathematician Nicholas of Linna, who had allegedly explored the Arctic around 1364, reported the presence at the geographical pole of a huge, black magnetic rock, at the crossroads of four in-drawing seas between four islands. The now-lost narrative may have served to further strengthen the belief that the source of geomagnetic attraction was to be sought on the Earth's surface, and most likely in polar regions.[26] Similar instances of a shift toward the north are easy to find in contemporary cartographical representations featuring a magnetic mountain or island there. Sixteenth-century examples include work by Johann Ruysch (world map), Olaus Magnus (map), Gerard and Rumold Mercator (globe, world map, and atlas), John Dee (map), Willem Barents (polar chart), Cornelius Wytfliet (map), and Philip Apian (globe).[27]

The sixteenth century is a veritable cauldron of different geomagnetic ideas, thrown together in conflicting and mutually adapting arrangements. Various celestial dipoles (both axial and tilted) competed for the scholar's favor with the legendary mountain. Meanwhile, Iberian and French navigators developed over a dozen geomagnetic longitude schemes along more practical lines (to be discussed below). Even though the level of sophistication of all postulates was still relatively low, the diversity has never been greater than at this time. This was probably caused by the advances in astronomy, cartography, and mathematics—yet unmatched by a large and widely available body of reliable geomagnetic data by which to build, test, and falsify hypotheses. Overseas trade empires were only just emerging, and instruments left much to be desired in accuracy and standardization. Nevertheless, the few geomagnetic observations that were gathered by natural philosophers seemed to suggest that the earliest solutions were too simplistic to fit reality. The magnetic mountain was in this respect far more flexible than the silently revolving dome overhead; a large rock of lodestone could easily be situated elsewhere, should theory or measurements require it, whereas a star was deemed to hold its (relative) position in the sky forever. This greater adaptability explains the numerical preponderance of crustal over celestial ideas in the second half of the sixteenth century. At this time, the concept of the magnetic mountain also evolved, shedding its legendary overtones in favor of a more rigorous, mathematical approach that concentrated on a positional fix of the actual point of attraction. The work of Gerard Mercator provides a fine example.

The earliest intimation of a crustal explication of the Earth's field by this famous cartographer dates back to 1541, when he made a globe depicting an imaginary magnetic island north of Siberia. It did not seem to occupy a carefully calculated spot, but appears to have been merely placed in high northern latitudes. Six years later, in a letter to his patron, Mercator set out his thoughts on the matter more clearly. Not only did he explicitly put his confidence in a magnetic pole on Earth, but he also laid down his first attempt to determine its coordinates. By means of a spherical cross-bearing, he found the intersection of the great circles of declination through Flushing on the Dutch coast and Danzig (Gdansk) in Poland to lie at a distance of about 11° colatitude from the geographical pole (see fig. 2.3). Its longitude (168° E) he still reckoned from the Canaries, following the tradition inaugurated by Ptolemy.[28]

Twenty-eight years later, Mercator's world map (1569), drawn on the famous "Mercator projection," had a resounding impact. It is quite remarkable in hinging on two mutually exclusive postulates, based on the choice of prime meridian. Rather than Ptolemy's prime meridian over the Canaries, the baseline to reckon east and west from now coincided with the Atlantic agonic (presumed to be meridional, resulting from a tilted dipole). Confusion arose after conflicting reports situated observed zero declination both in the vicinity of

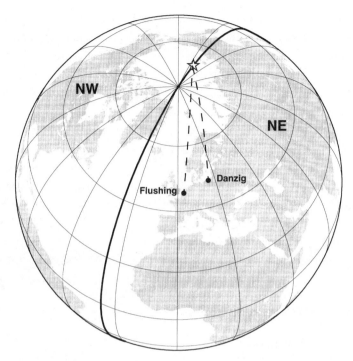

Fig. 2.3. Mercator's 1547 attempt to locate the magnetic mountain at 79° N 168° E (marked by a star in the figure) by spherical cross-bearing (dotted lines). Declination values for Flushing and Danzig yielded an agonic great circle (solid line), dividing the globe in two equal areas of northeasting (NE) and northwesting (NW) respectively; the discrepancy between the mountain's coordinates and the intersection results from inaccuracies in Mercator's calculation. Orthographic projection.

Corvo (Azores) and several of the Cape Verde islands. Rather than committing himself to a single interpretation that might prove to be wrong, the cartographer decided to calculate the intersections of each prime meridian (when extended through 180° east) with the great circle through Regensburg (in Bavaria), where declination then reportedly amounted to 16°44′ east of true north. Not surprisingly, this exercise yielded two prospective sites for the magnetic island, which were marked separately on the map.

Given that the field was still considered time-invariant, it seems strange that Mercator did not incorporate his previously compiled observations at Flushing and Danzig in his new calculation, to obtain a positional average. Perhaps the inconsistency of the two reported locations of the agonic meridian had appeared irreconcilable to him. Had he worked from the assumption that agonic lines (and isogonics in general) could be curved, none of the observations need necessarily have been in conflict. But that would have meant giving up the pleasantly uncomplicated antipodal dipole. It seems doubtful that he would have even considered this option, at a time when the notion and consequences of a geomagnetic south pole had barely sunk in. Thus, while Mercator's ideas present a good example of a more mathematical approach and the novel discomfort caused by conflicting data, its author was nevertheless still a member of the "old school," in implicitly adhering to the dipole concept while concentrating solely on the northern hemisphere. Others would soon widen this limited focus in reappraisals of the physical manifestation of the perceived central point of attraction. As a result, new images of magnetic islands and mountains became very scarce in the seventeenth century. An isolated instance, in a book on precious stones by physician Anselme Boëce de Boodt (in 1609), reported a sighting of a mountain at colatitude 17°, 180° east of the Cape Verdes. In the French edition of 1644, the publisher added the soothing note that the mountain did not constitute a danger to shipping, as its power was insufficient to tear vessels apart.[29]

Three Iberian Ideas

Most geomagnetic hypotheses presented thus far have at least two aspects in common: some traces of a classical origin, and their creators' lack of direct experience of oceanic navigation. The converse is true for systems developed in Portugal and Spain during the sixteenth and early seventeenth centuries. Many proponents there had personally guided ships to distant shores, or were otherwise closely associated with the daily practical concerns of the navigator. Iberian mariners were the first Europeans to turn the trackless oceans into a network of sea-lanes, while confronting the limitations of dead-reckoning and plane chart. To them, the problem of determining longitude was particularly acute, not just because of seafaring considerations, but also due to territorial disputes. On 4 May 1493, a papal bull had divided the world in two, along a meridian 100 leagues west of the Azores. Spain was entitled to pursue its interests in the western hemisphere, while Portugal obtained the rights to trade goods and found colonies in the other half.[30] After vehement protests by the Portuguese against the unfairness of the division, the Treaty of Tordesillas a year later adopted an alternative demarcation 370 leagues west of the Cape Verde Islands. A new controversy subsequently arose over possession of the Spice Islands (the Moluccas) and the Philippines, reckoned by each party to lie just within its own hemisphere of influence. In the ensuing years, cosmographers and other experts from both countries conducted research and conferences to try to resolve such matters as the actual location of the prime meridian and the length of a league, without success.[31] Geomagnetic arguments doubtless played an important part, especially in the placement of the line of division.

The area centering on the triangle between the Azores, the Cape Verdes, and the Iberian Peninsula was already a focal point of maritime activity at the turn of the sixteenth century, and it was from this region that a number of consistent observations of magnetic declination was gathered. These laid the foundations for choosing an agonic prime meridian, and for the development of the very first tilted dipole concept in 1508, decades before it gained a foothold in northwest Europe. Its Portuguese protagonists were João de Lisboa and Pedro Anes, two navigators who collaborated on the analysis of magnetic data. The system came out in print six years later, in de Lisboa's *Tratado da Agulha de Marear*. In plain, factual language, and based on observations along the track between Lisbon and the Azores, the author proposed a direct relationship between magnetic declination and longitude. It assumed that declination would alter one compass point ($11°15'$) for every 250 leagues sailed east or west at a certain latitude. The expected global pattern of a tilted dipole would result in northeasting to east, and northwesting to west of the agonic prime meridian, which itself was mirrored on the antimeridian at $180°$ of longitude. Halfway in between, at $90°$ east and west, the needle would show a maximum of four points ($45°$) of magnetic declination (see fig. 2.4).

De Lisboa's description was wanting in one important aspect, namely the colatitude of the dipole. Arguably, this was not of any practical concern to the mariner, since few would venture further from the equator than 65 degrees anyway (the highest latitude listed in the accompanying table). Regrettably, it is impossible to reconstruct the dipole's tilt in hindsight. The whole notion of such a system having meridional isogonics other than zero is mathematically flawed; an imaginary observer on a dipole world, traveling north along a meridian other than the agonic would find that the angle between true and magnetic north continually increases. This underlines the more practical approach taken, concentrating on a description of the observed and predicted local pattern, rather than a global representation. This geomagnetic solution to the longitude problem thus contains the seeds of two different ideas: the (global) tilted dipole and the (local) declination-distance relationship. Both were to be pursued by others during the sixteenth and early seventeenth centuries.

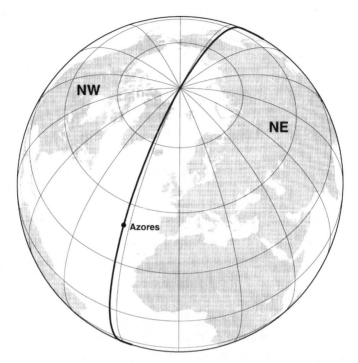

Fig. 2.4. De Lisboa's tilted dipole (1508), at unspecified latitude and 180° E, based on zero declination at the Azores, yielded an agonic great circle (solid line), dividing the globe into two equal areas of northeasting and northwesting respectively. Orthographic projection.

The Portuguese Faleiro brothers, Ruy and Francisco, worked as navigators in Castilian service. Together they wrote a work on navigation, which the explorer Magellan brought along on his circumnavigation in 1519–22. Much later (in 1535), Francisco Faleiro reworked the manuscript into a printed edition as *Tratado del Esphera y del Arte de Marear*. In 1566 he also became involved in the Tordesillas Treaty dispute, when together with Pedro de Medina, Alonso de Santa Cruz, and other cosmographers, he gave evidence before a royal council on the longitude of the Philippines.[32] The tilted dipole system sketched in the treatise was less detailed than de Lisboa's. It featured an agonic meridian over Corvo, the dipole in the northern hemisphere at 180° east and unspecified co-latitude, and maximum declination at 90° east and west longitude, without further particulars. Similar postulates lacking information on the dipole's co-latitude appeared in the ensuing decades, put forward by Spaniards Alonso de

Santa Cruz (1542), Gonzalo Fernandez de Oviedo (1547), and Pe(d)ro Menéndez de Avilés (1573).[33]

The scheme of de Santa Cruz formed part of his royal survey of longitude proposals, and assumed a regular increase of declination with longitude up to a maximum of three compass points (33°45′), rather than de Lisboa's four points. Menéndez de Avilés in turn rejected three points (without explicitly mentioning de Santa Cruz) in favor of a peak value of fifteen degrees. At the time, it was a mere rhetorical gesture, since de Santa Cruz had himself already abandoned his own hypothesis three years after publication. That decision had been based upon the first extensive compilation of magnetic declination measurements made by Portuguese explorer João de Castro, which seemed to contradict the earlier-postulated even dispersion along meridians. De Santa Cruz is thus quite exceptional in rigorously applying the principle of falsification to the fruits of his own labor, whereas some other proponents of longitude schemes suffered from Pygmalion syndrome: falling in love with their work, and stubbornly defending it as the best possible representation of reality, in spite of substantial evidence to the contrary.

A more fleshed-out tilted dipole, including colatitude, saw the light of day thanks to Spanish *piloto-mayor* Pedro de Syria in 1602. It was reminiscent of Cortés's celestial system and is chiefly noteworthy for still displaying uncertainty as to whether the pole was to be found on Earth or in the heavens.[34] Around this time, two other types of interpretation had won favor on the Iberian Peninsula. The first was the distance-declination ratio (see above), the second the static quadrupole (see below). Regarding the former, it is peculiar to find an interval as large as a hundred years between its introduction by de Lisboa in 1508 and further elaboration by compatriots from 1608 onwards, and even more so when French developments along these lines are taken into account. These occupy a single isolated period from 1542 to 1583, and warrant a short discussion.

Dieppe hydrographer Jean Rotz is chiefly known for having invented the *cadrant differential*, a complicated instrument combining sundial, compass, and astronomer's rings. Once the instrument was set for the correct latitude, a wire would cast a shadow, enabling true and magnetic north to be read off simultaneously. In 1542 Rotz crossed the Channel and offered the device, a manuscript treatise on nautical science, and his services to England's King Henry VIII, who granted him residence and a post as Royal Hydrographer. Rotz's *Manuel de Hidrographie* assumed a meridional agonic to lie over the isle of Ferro (Canaries). Declination readings at four distant locations around the Atlantic Ocean he subsequently combined with their (tentative) distances from the prime meridian to obtain a single average of one degree of declination for every 22½ leagues sailed east or west at any latitude. The selection of observations (from the author's own experience and other French voyages) suggest that de Lisboa's earlier concept probably did not serve as an inspiration. Furthermore, the relationship between declination and distance sailed along a parallel was independent of latitude in Rotz's scheme.[35]

By contrast, the *Dialogue de la Longitude Est-Ouest,* written by Normandic author Toussaints de Bessard thirty-two years later, did include a table of latitude-dependent correction factors by which the longitude could be found for various distances from the equator. The author assured his audience that the whole was founded on "observations made in diverse countries."[36] At the same time, the method tried to draw on astronomy for additional support, claiming that the magnetic force originated in the pole of the ecliptic. The fact that local declination would then have performed a daily and a yearly cycle he overlooked. Perhaps this was why naval pilot Jacques de Vaulx in 1583 omitted this part when he copied some of Bessard's work. De Vaulx moved his prime meridian one degree to the west of the latter's, and replaced the latitudinal correction table by a constant factor of two.[37]

Two Iberian heirs to de Lisboa's legacy claimed ownership of their inheritance early in the seventeenth century. They were Luis de Fonseca, from Portugal, and the Spaniard Lorenzo Ferrer Maldonado. Both submitted elaborate proposals of a distance-declination solution to the longitude problem to the Spanish crown. The bureaucracy in processing the various applications for the reward offered by Philip III has ensured that at least some remnants of the relevant characteristics still reside in the archives of Seville's "Trade House" (Casa de Contratación). De Fonseca's correspondence and other papers span the period 1608–13, and mostly relate to requests for financial support. They serve to illustrate that protagonists hawking a longitude scheme were in a difficult position. On the one hand, they had to engender as much interest and confidence in their conjectures as possible, while at the same time withholding crucial details until such time as the maximum possible amount of money had been received, either in the form of travel grants (for magnetic surveys or personal attendance at court), as aid in building instruments and performing experiments, or as part of the longitude reward itself.

In de Fonseca's case, the information missing from the written exchanges is his secret technique of needle magnetization, which he claimed would cause it to decline regularly with longitudinal displacement, regardless of latitude. A meridional agonic over the Azores furthermore suggests an underlying tilted dipole concept. The heart of his scheme was a special compass, the needle of which could be reset to zero at an arbitrary point of departure. Traveling east or west would then supposedly result in a commensurate needle displacement in the same direction. Instead of supplying a table to convert degrees to distances, the graduation directly expressed local declination in leagues from the meridian of departure. Navigators eventually tested the system on land along the triangular route Madrid-Seville-Lisbon-Madrid, and at sea on a voyage to the West Indies. Despite alleged favorable results and some piecemeal royal support to investigate matters further, nothing appears to have come of it.[38] A few years later, in 1615–16, captain Maldonado tried his luck with a somewhat simplified version, assuming one degree of longitudinal change for one degree of declination. His implied notions of a tilted dipole are similarly vague (an agonic prime meridian over Corvo, without specified colatitude), but he did state that everywhere on Earth the needle directly respected the (magnetic) pole.[39]

Neither of these two proposals acknowledged either the efforts of de Lisboa or those of their three French predecessors. This fact need not raise eyebrows. The idea itself is straightforward enough not to require a solid foundation built by generations of scholars; moreover, the observations were few, they covered a limited area, and each proponent collected his own. The ideas grew out of the most direct relationship imaginable between magnetic measurements and seafaring along a parallel of latitude, and numerous navigators may have stumbled upon it at some point in their careers. A larger body of reliable data would, however, have quickly dispelled such extreme reduction of global patterns to a single rule. Their appearance in the form of a serious proposal therefore indicates a certain level of (localized) awareness of the field's complexity. These schemes developed and were superseded within a relatively short period of time, between the recognition of declination itself, and a "critical mass" of available observations conflicting with theory, stimulating the search for more intricate alternatives. Spain reached this stage in the 1590s, and this gave rise to a third family of representations, based upon the realization that two magnetic poles were not sufficient to explain the observed field characteristics.

From the 1570s onward, an increasing number of Portuguese navigators had started to publish *roteiros,* sailing directions containing dozens of compass readings, personally collected at way stations and destinations. Among them was Vicente Rodrigues, who had identified at least three places where magnetic

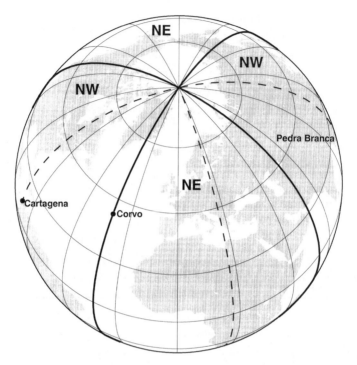

Fig. 2.5. Da Costa (1596) assumed four agonic meridians (dotted lines) to be placed at equal intervals of 90° longitude, dividing the globe into four equal sections (solid lines) of northeasting and northwesting, relative to Corvo (Azores). Orthographic projection.

and true north coincided: at the Azores, at Cape Agulhas (South Africa), and at Pedra Branca (close to either Singapore or Canton in China).[40] Spanish exploration of the Americas added a fourth, where the West Atlantic region of northwesting gave way to northeasting, which reigned throughout most of the Pacific basin. "A very able Portuguese pilot told me," wrote José de Acosta in 1590, "that there are four points in the whole world where the needle is fixed with the north."[41] This Spanish Jesuit did not specify a particular place associated with this fourth instance of zero declination, although he had been a missionary in Peru, and thus may have been able to obtain firsthand information. In a slightly later hydrographical and navigational treatise written in 1596, a Spanish professor of mathematics, Francisco da Costa, designated Cartagena (Colombia) as situated on this fourth agonic line. As with the other three isolines, he still imagined it to run meridionally from pole to pole.[42] Poor determination of the longitudes of these places made it possible to estimate the distance between each pair of adjacent agonic meridians to be exactly 90° of longitude. Thus, reckoning from the Azores, Cape Agulhas would lie at 90° east, Pedra Branca at 180°, and Cartagena at 270°. In other words, two perpendicular great circles quartered the globe in four equal parts of alternating sign of declination (see fig. 2.5). Needle displacement would furthermore reach a maximum on the meridian in the middle of each sector. This constitutes a quadrupole system, formed by two antipodal dipoles.

Nearly three decades later, Portuguese cosmographers Valentim de Saa (1624) and Manuel de Figueiredo (1625) added more particulars.[43] The declination was said to reach a peak value of two compass points (22°30′), Cartagena had become Vilalobos on the Pacific coast (west of Acapulco, Mexico), and the positions of the agonic meridians over Corvo and Cape Agulhas had moved further apart. De Saa moreover mentioned Vicente Rodrigues and the experiences of modern pilots as his sources, and offered a table of their findings. His version (with Pedra Branca and Mexico hosting agonics) also formed the basis for the longitude scheme proposed to Philip III by Cristovao Bruno, an Italian Jesuit who taught navigation in Portugal during the 1620s.[44] This particular scheme was remarkable in being laid down not only in writing, but also in graphical form, most likely the earliest isogonic chart ever produced. Regrettably, neither has survived.

The year after de Saa's publication, Figueiredo replaced Pedra Branca by the city of Canton. It remains unknown whether he supposed magnetic declination to be latitude dependent. Fifteen years later, a certain Nicolas Le Bon translated Figueiredo's tract into French (1640), and also added some of his own conjectures. In his version, Pedra Branca was back in favor over Canton, and the two great circles had been raised from their Earthly domain to lie outside the celestial globe.[45] In an ill-fated attempt to reintroduce an astronomical explanation, he postulated their two intersections to coincide with the poles of the ecliptic, at 23°30′ of arc distant from the geographical poles. Following this fixed colatitude, the author confidently supported the notion of latitude-dependent declination between the agonic demarcations. Unfortunately, Le Bon's hybrid did not make much mathematical sense, as he somehow expected a single dipole to produce two agonic great circles perpendicular to each other. The choice of the zodiac poles was similarly awkward, since (as was stated earlier) they appear to perform a daily and a yearly cycle relative to the Earth's surface while the planet rotates and orbits the Sun.

Another fifteen years on (1655), Portuguese cosmographer Antonio de

Mariz Carneiro gave the quadrupole a new lease on life. He revived it only in a rudimentary form, keeping the perpendicular great circle agonics and the two-point maxima, but omitting all particulars regarding their longitudes. An interesting detail is his continued reliance on data procured by Vicente Rodrigues, by that time almost a century out of date. It is therefore likely that the author was either unaware of, or unconvinced of, the existence of the field's change over time (by then discovered). The text was copied almost word for word (and without acknowledgment) by Luiz Serrão Pimentel in 1673, with a second edition following eight years later. By that time, the system had become as stale as the observations it had initially been based upon.[46]

Dutch Derivatives

Clear evidence exists of Portuguese navigational knowledge ending up in the Northern Netherlands during the last decade of the sixteenth century. At the time, the Dutch were about to start their highly profitable ventures in the East Indies, first through a number of separate enterprises, and then, from 1602 onward, unified in the Dutch East India Company, the VOC (Verenigde Oost-indische Compagnie). But Dutch shipping had until then mainly traversed European waters, and the authorities recognized the need for reliable intelligence regarding routes to the East. As the Portuguese were understandably wary of potential competitors, information had to be procured in secret. A pivotal role in this respect was played by Jan Huyghen van Linschoten of Enkhuizen. After traveling to Portugal in 1579, he managed to find employment with the Bishop of Goa (India). During many years of voyaging in Asia, he amassed a sizeable collection of roteiros.

Upon his return to his hometown in 1592, he set to work on translating these sailing instructions, and eventually published them in three volumes. The last part rolled off the press just in time to be taken along on the first Dutch voyage to the East (1595–97). It contained excerpts of the "Carreira da India" by Diego Affonso, as well as several manuscript sailing directions by Vicente Rodrigues. These instructions list roughly the same set of declination measurements as those used by the Portuguese and Spanish to construct quadrupole systems. Meanwhile, the brothers de Houtman, future members of the expedition, spent more than a year in Portugal on a secret mission to secure additional manuscript sailing directions. In 1594, van Linschoten moreover acquired a copy of de Acosta's *Historia Natural y Moral de las Indias* (1590), with

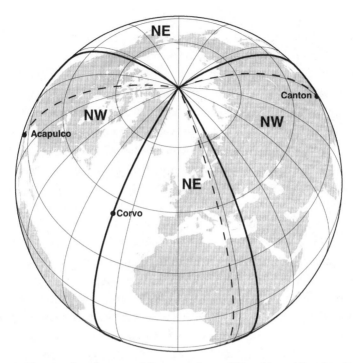

Fig. 2.6. Plancius (1598) assumed four agonic meridians (dotted lines) to be placed at intervals of 60° and three times 100° of longitude, dividing the globe into four unequal sections (solid lines) of northeasting and northwesting, relative to Corvo. Orthographic projection.

the aforementioned reference to four places of zero declination. A Dutch translation followed in 1598.[47]

A driving force behind the foundation of the first Dutch East Indian trade companies was the Calvinist preacher Petrus Plancius (of Flemish origin), who was closely involved with the preparations for the undertaking. His experience as a cartographer served him in good stead when he drafted his own sailing directions, which included many values of observed declination. Whereas van Linschoten had primarily incorporated Portuguese sources, Plancius relied more upon Spanish material. Based on these data, in combination with the ship's logbooks handed in after the explorers' successful return in 1597, the versatile minister produced two new manuscripts for the second Asian adventure the following year. One was a revision of his earlier directions based on the recorded experiences of the masters and mates, the other a practical instruction to find longitude by means of the geomagnetic field.[48]

His concept posited a (mathematically inconsistent) quadrupole on the Arctic Circle (at colatitude 23°30′), of which the agonics did not form great circles, but only meridians, at 0°, 60°, 160°, and 260° longitude east. Relative to the chosen prime meridian of Corvo, these were supposed to pass over Cape Agulhas (actually at 51°06′ E of Corvo), Canton (at 144°21′ E), and Acapulco (at 291°10′ E), a familiar foursome. Note that de Saa and Figueiredo (about a quarter century later) stuck to a symmetrical arrangement forcing a ninety-degree interval; Plancius had neared a lot closer to the actual longitudes of these places (see fig. 2.6). It is furthermore of interest that, in stark contrast to modern analysis, measurements of (zero) declination were taken at face value, whereas their geographical location was still prone to considerable uncertainty.

The Plancius agonics once again separated sections of opposite needle deflection, peaking midway at 30°, 110°, 210°, and 310° E longitude. Declination was furthermore deemed latitude dependent and differed from all previous quadrupoles in having minimal values along the parallel at 20° north, rather than on the equator. At the end of his tract, the author had listed forty-three observations to corroborate his interpretation, which, however, were restricted to the Atlantic hemisphere, and probably had their longitudes "adjusted" to fit the conjecture.[49] Most remarkable of all, for about three decades some Dutch mariners implemented this scheme at sea to find longitude. As the spherical trigonometry was rather involved, a specially developed instrument, the "longitude-finder" (*lengtewijser*), solved the problem by construction instead of calculation. Traces of the method in logbooks are unmistakable, albeit scarce. Although Plancius had taught navigators the geomagnetic principles on a few occasions, there does not seem to have existed any directive, either by the Dutch East India Company or its predecessors, for masters and mates to abide by his maxims. Some applied the technique, others did not. Of those that did, some were favorably impressed, while others condemned it in the strongest possible terms.[50] Criticism seems to have mounted over the years, though, and not only from mariners. Cartographers Jodocus Hondius (1597) and Willem Jansz Blaeu (1608), hydrographer Albert Haeyen (1600), mathematics teacher Robbert Robbertsz (1612), and professor of mathematics Adriaen Metius (1614) all dipped their pens in acid when they rejected the idea.[51]

The Plancius hypothesis was revised twice, both times by Dutch polymath Simon Stevin. In 1599, he published a short pamphlet called *The Haven-Finding*, which contained:

— the table of observations compiled by Plancius (with acknowledgment);
— a method to find the *direction* of the port of destination (the "haven-finding art" of the title);
— a hypothetical conjecture based on the Plancius system, but mathematically consistent.

The author also rejected the thought of a magnetic pole attracting the compass needle, and gave instructions on how to measure declination at sea. In all he wrote on the subject, he was careful to stress the preliminary nature of the postulates, subject to revision when new and more reliable data would become available. His practical method was nevertheless so sensible that it would have continued to work regardless of changes in the field's local characteristics. It basically consisted of an adaptation of latitude sailing, especially suited to find small islands. When the ship arrived at the latitude of the intended port of call, the known declination at the destination was to be compared with the value observed on board. The distribution pattern in the area, as well as progressive readings on subsequent days, would then indicate whether one should sail due east or west to reach harbor. The technique was not trumpeted as the ultimate solution to the longitude problem, but merely offered as a modest aid to navigation. Furthermore, it has to be distinguished from Stevin's amendment of the Plancius concept, a separate, theoretical construct without practical application. Here, the author proposed to dispense with the Mexican agonic in favor of extending the other three to form great circles. He thereby created the first recorded instance of a sextupole proposition (see fig. 2.7). Six meridional agonics thus sectioned the globe in unequal parts, at 0°, 60°, 160°, 180°, 240°, and 340° longitude. The alternating sign when crossing an agonic line was the only conclusion drawn regarding a predictable dependence of declination on longitude within a sector.

A second revision followed nine years thereafter, when *The Haven-Finding* became part of Stevin's great work "Mathematical Thoughts" (*Wisconstighe Gedachtenissen*). By this time, he had dropped all postulates pertaining to longitudes greater than 160° east. In addition, the mathematician explicitly distinguished between longitude-finding and haven-finding. Elsewhere, he expressed his newfound close alignment with a different opinion of the geomagnetic field, promulgated by William Gilbert in 1600 (of whom more will be said later), which left more flexibility in the location and shape of agonics in general.[52]

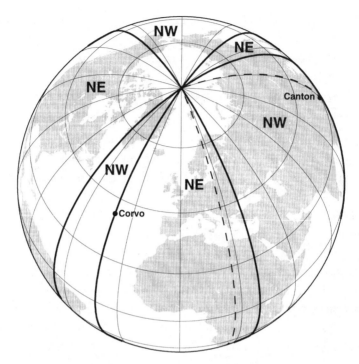

Fig. 2.7. Based on Plancius's data, Stevin (1599) assumed three agonic great circles to divide the globe into six unequal sections (solid lines) of northeasting and northwesting, relative to Corvo. Orthographic projection.

Whereas Plancius had a more direct impact on Dutch navigation to the East Indies in the early seventeenth century, Stevin was more widely read in Europe. In 1599, the year of the first publication, a Latin translation of *The Haven-Finding* also appeared, which made its contents accessible to many European scholars. Meanwhile, in the same year lecturer and mathematician Edward Wright supplied an English version, which he also incorporated in several editions of his famous treatise *Certaine Errors of Navigation*. A Latin publication of Stevin's work on mathematics was undertaken by Willebrord Snellius, while a French translation appeared in 1620.[53]

The rapid spread of these texts assured dispersal in neighboring countries of the Plancius table of observations and Stevin's remarks on geomagnetic navigation. The haven-finding art agreed well with the developing practice on board ship to use magnetic declination as a positional clue. The problem with the tabulated values of Plancius was that the entries were undated, and confidence in them plummeted as the discrepancies with the changing field grew over the years. Even when secular variation was eventually recognized, the missing temporal dimension prevented them from reappraisal for time-dependent analysis (both in the past and at present), which is most unfortunate.

Dip and Latitude

Instead of trying to balance the needle in the horizontal plane to measure declination, one could alternatively pivot the needle on a level axis to measure the direction in the vertical plane. Inclination generally came to the fore the farther one traveled from the magnetic equator, pulling one end of the compass needle downward as one neared the poles. Nevertheless, isoclines do not form a regular distribution like geographical parallels; their curved patterns display too much variation to be able to replace celestial information for determining latitude. But without isoclinic charts, this conclusion is hard to draw, as several dip-latitude schemes from around the turn of the seventeenth century attest. "The learned in our age . . . have beaten their brains so much, for many months and years together, on the hypothesis of the magnetical inclinatory needle,"[54] wrote longitude-finder Samuel Fyler in 1699, not without some glee, looking back upon the vain efforts to gain positional information from the angle of the vertical needle. Nevertheless, it was in part through the pursuit of this dream that inclinometers for shipboard use were developed in late-sixteenth- and early-seventeenth-century England. Examples are featured in a book by mathematician Thomas Blundeville, who devoted an appendix of his 1602 *Theoriques of the Seuen Planets* to the description and application of such instruments. A broad hoop was made of brass and divided into four quarters, each graduated in ninety degrees. Two thin plates were then fixed horizontally, and between them held the axis on which a magnetized needle of a little under five inches was to turn in the vertical plane. Both sides then covered with circular panes of glass, it was to be hand held by a suspension ring, and oriented in the magnetic meridian. Local inclination could then simply be read off. Other designs had the instrument fixed on a stand or gimbaled in a box.[55]

Investigations of magnetic inclination furthermore strengthened the growing conviction that the origin of the Earth's field resided somewhere deep *inside* the planet, instead of upon the surface, or above it. Research was primarily conducted on land, mostly in England, and based on a remarkably poor record of observations. Whereas sailing directions, ships' logs, and the first compilations together made about a hundred measurements of declination available for scholarly study at this time, the first dip schemes were based on a single inclination value of 71°50' (from the horizontal) at London. It was obtained by compass maker Robert Norman, shortly before his publication *The Newe Attractiue* in 1581. In this tract, he rejected the notion of a crustal pole as envisaged in the legend of the magnetic mountain, opting instead for a point far below the surface that all magnetized needles would "respect": "This straight lyne must be imagined to proceede from the center of the needle into the globe of the Earth, extending, and going directly forth, both wayes infinitely. But in what part of this line the point respectiue is . . . we must leave untill the expert travailer have made certaine observation of this declyning of the needle in other places."[56]

Although the pole's exact whereabouts remained to be discovered, Norman still believed that inclination was governed by this solitary, fixed point, which implied a means with which to divine latitude at sea. Around the same time, fellow experimenter in magnetism William Barlowe improved upon the design of the dip meter, and likewise avowed his confidence in this method; he considered it especially useful when the skies remained overcast for long periods. But neither in his 1597 publication nor in later writings did he commit himself to particulars. It was left to Henry Briggs, Gresham professor of geometry, to express inclination as a function of latitude in the late 1590s. A number of writers subsequently reproduced this idea with alterations, in the form of a simple dip-latitude table.[57]

A different tack was taken by French nobleman and Royal Geographer Guillaume de Nautonier. The tilted dipole system he proposed in his book *Mecometrie de l'Eymant* in 1603–4 stood at an angle of 23° from the rotation axis (note the proximity of the Arctic Circle), and 180° east of the westernmost Canary Isles. In addition to a huge table of predicted declinations all over the world, a dip-latitude scheme was also thrown in for good measure. At first sight, it may have seemed little different from the ones encountered earlier, until one realized that the inclinations were related to *geomagnetic* latitude, rather than geographical latitude. De Nautonier had superimposed a system of parallels of equal inclination at a tilt of 23° on the geographical grid, thereby killing two birds with one stone: given either longitude or latitude, the other half of the coordinate pair could be derived from the table.[58]

Soon after publication, de Nautonier offered his vision to the French king Henry IV in the hope of receiving a reward. Regrettably, mathematical consistency offers no guarantee of any bearing on reality. The aspiring claimant seems to have been somewhat selective in admitting evidence, allowing only those observations that could bolster his supposition, and rejecting all other data. The differences between the concept and the real world soon became apparent through the work of a Paris professor of mathematics, Didier Dounot de Barleduc, who took the whole concept apart in a devastatingly critical appraisal, listing unfounded assumptions, errors in calculation, and data manipulation. Nevertheless, de Nautonier's son Philippe came to England in 1654 and tried, without success, to sell his father's idea to Oliver Cromwell.[59]

Despite the ardent convictions of these latitude-finders, the measurement of inclination never attained a place in standard navigational practice, and no logbook can testify to its implementation to derive latitude. Scattered sets of readings have primarily been compiled at sea by some explorers, natural philosophers, and the odd naval survey.[60] Two reasons come to mind as to why the dip was so little regarded. Firstly, the idea of magnetic latitude was quickly abandoned, being less reliable and precise than celestial information. Secondly, the measurement itself was more difficult than that of declination, the instrument having to be kept both level with the horizon and precisely aligned with the magnetic meridian at all times. On a constantly moving ship, this required a regular compass nearby, which in turn allowed deviation to come into play, the two magnetic needles affecting each other's orientation. The navigator already had enough to keep him occupied without having to deal with these additional concerns, from which no practical benefit would accrue.

In looking back upon the sixteenth century, a few characteristic developments stand out. First and foremost, the acknowledgment of declination as a real phenomenon instigated the abandonment of axial dipole concepts (phase one) in favor of tilted dipoles (phase two). Once the "natural" link with the celestial (or geographic) pole was severed, the actual placement of the attractive point became prone to many interpretations. In the heavens, the Polestar and the pole of the ecliptic were considered, while on Earth the magnetic mountain conjecture eventually favored the more mathematical method of localization by cross-bearing. Both approaches still merely considered the northern hemisphere, although they implicitly represented the global field.

Navigational practice spawned two other types of hypotheses. One worked from the assumption of a direct relationship between longitudinal distance and magnetic declination, while the other used the measured positions of (meridional) agonics to construct tilted dipoles and quadrupoles, with explicit north and south magnetic poles. The declination-distance relationship seems to have spontaneously regenerated at different places without exchanges of ideas or data. Contrastingly, agonic schemes were clearly subject to mutation by various individuals, who amended earlier attempts by altering the parameters. Evidence of communication exists both on the level of data (the tables of Rodrigues, de Castro, and Plancius) and postulates (explicit references to predecessors). Language and limited circulation of publications may still have been a barrier, restricting the diffusion of some ideas to compatriots. De Acosta's quadrupole was echoed most often on the Iberian Peninsula, de Vaulx's work contains traces of fellow Frenchmen Toussaints de Bessard and (possibly) Rotz, and Stevin was the first scholar to publicize the Dutch Plancius table. On the other hand, the quadrupole was later copied in France (by Le Bon), and may very well have inspired Plancius as well. In addition, Stevin's *The Haven-Finding* soon found its way to England and France.

An important question is to what extent the dynamics of theoretical developments were driven by the observations. In the sixteenth century, poor charts (resulting in longitudinal uncertainty), a lack of standardization in instruments and measurement, as well as the numerical paucity of geomagnetic data (mostly concentrated in small pockets scattered across the Atlantic hemisphere) all worked against the rigorous testing of any hypothesis. Nevertheless, most of these bore the hallmark of incorporated observations: the distance-declination ratio, the selection of the proper agonic prime meridian, de Santa Cruz's rejection of his own proposed longitude scheme upon learning of de Castro's findings, the falsification of the Plancius method by testing at sea, the rejection of de Nautonier's by other empirical evidence, and, upon the discovery of inclination, the realization that the dipole should be situated in the Earth's interior rather than in the crust. Note, however, that this does not preclude preselection of "supportive" evidence on the part of the proponents. It remains impossible in hindsight to assess for each individual case what declination values may have been seen but discarded on improper grounds. What is certain, though, is that the error margins on the measurements and their longitudes were still wide enough to initially and tentatively confirm the wide variety of phase-two field representations. New insights would change all this. These ideas, and the clashes they gave rise to in natural philosophy, are part of the seventeenth century, to which the following chapter is devoted.

CHAPTER 3

The Age of Discord

Geomagnetism in the Seventeenth Century

It would be both easy and misleading to characterize seventeenth-century Europe solely in terms of political and religious strife. Admittedly, the Dutch War of Independence (1566–1648), the Thirty Years' War (1618–48), the English Civil War (1642–48) and Glorious Revolution (1688), the three Anglo-Dutch Wars (1652–74), and Louis XIV's Dutch War (1672–78) are but some of the century's grim events that made the history books. Meanwhile, emergent Dutch and English sea power successfully challenged the former Iberian partition of the world on many a faraway shore. American, African, and Asian colonies mushroomed in an increasingly intricate network of trade, while old spice monopolies gave way to new ones, and many of the world's coastlines became engraved in copper plate, where they had only been hand sketched before. Early in the century, the two future molochs of East Indian commerce came into being: the English East India Company received its charter in 1600; the Dutch Verenigde Oostindische Compagnie (VOC) followed two years later. Northwest European navies, when not engaged in battle, patrolled the Atlantic, the Mediterranean, and the home stations. Meanwhile, fruitless attempts to find Arctic routes to the East Indies had led to the discovery and subsequent exploitation of vast fishing and whaling grounds in the North Atlantic. On the other side of the globe, Australia became the latest identified continent in 1642.

But this was also the time when the writings of Francis Bacon and René Descartes announced the beginnings of a more scientific civilization, when Galileo Galilei published his *Dialogo* (1632) and Isaac Newton his *Principia* (1687), and when the first scientific societies started circulating their periodicals (1660s). This era celebrated the invention of the telescope and the microscope, the widespread adoption of decimals and logarithms, and the advent of statistics and calculus. The century also witnessed the discovery of Jupiter's moons and Saturn's ring, and the hard-fought acceptance of planets revolving in elliptical orbits around the Sun, rather than the Earth residing immobile at the center of nine shells that carried the universe in a divine "harmony of spheres." In this heady atmosphere of *rationalism* (theories) and *empiricism* (observations), many mutually exclusive interpretations of the natural world vied for dominance. Their influence on geomagnetic thought is evident throughout the seventeenth century, and provides a useful framework with which to interpret its development.

Gilbert's Magnetic Philosophy

In 1600 London physician William Gilbert published the results of some seventeen years of experimentation in his treatise *On the Magnet* (*De Magnete*).[1] It contained an introduction followed by five books, each of them devoted to one of the defined magnetic "movements": coition (the law of equality between action and reaction, replacing attraction), direction, declination, inclination, and revolution. Gilbert coined the famous phrase "the Earth itself is a great magnet," and has been called the founder of terrestrial magnetism. He furthermore investigated static electricity, the "orb of virtue" around lodestones of various shapes and sizes, and the magnetization of iron through shock, deformation, and changes in temperature. He performed his experiments systematically, in relation to an antecedent hypothesis to be proved or falsified. Yet to call him a true empiricist would be an overstatement. He adhered to the doctrine of the animate nature of the universe in general and of the lodestone in particular, and supposed that the emitted magnetic effluvia allowed a magnet to act at a distance by actuating the recipient's magnetic Form.[2] In his perception, magnetic motion was therefore not a contact phenomenon. In the case of the Earth (or any other celestial body), its matter gave it firmness, direction, and movement, while its "magnetic vigor" imparted aggregation of parts and "verticity," the self-directing capacity of the globe, maintaining its orientation. Gilbert believed the Earth's insides to be largely made of iron ore, while he interpreted lodestone at the surface as a stony concretion of the same material. He also believed that magnetic rocks possessed a soul, and saw them as the offspring of their living mother planet.[3]

This interpretation was particularly relevant in the case of the spherical lodestones that Gilbert used in much of his research. The idea of studying a terrella goes far beyond that of mere analogy. It constitutes the establishment of *mimesis* in seventeenth-century philosophical practice, that is, direct imitation rather than symbolic representation: the Earth *is* a magnet, and the terrella *is* a little Earth. Its coordinate system was analogous to that of the Earth,

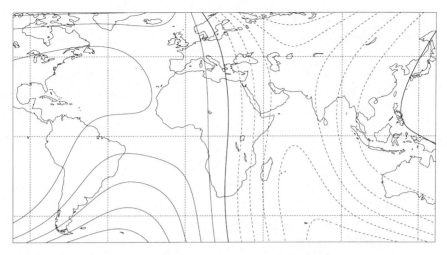

Fig. 3.1. Modeled magnetic declination at the Earth's surface in 1600. Bold curves = agonic lines; solid curves = northeasting; dotted curves = northwesting; contour interval = five degrees. Cylindrical equidistant projection.

and its nucleus was assumed to be the center of force, the magnetic energy being concentrated at the two poles. Gilbert rejected direct polar attraction, attributing needle orientation instead to a disposing influence exerted by the whole Earth. An explanation of magnetic declination was, however, more difficult to bring into line with the terrella's exhibited regular field pattern. Since the physician correctly deemed the deviating influence of iron mines to be superficial and very localized, he had to take recourse to a more global distorting agent. He found it in the distribution of continental landmasses.

Crustal magnetic matter on land, he argued, would affect a compass at sea, whereas the water and small islands would not. Experiments on a terrella with a gap (to represent an ocean) seemed to confirm this notion.[4] This had profound consequences for longitude schemes based on postulated symmetrical field line arrangements, as Gilbert wrote: "Variation is in divers ways ever uncertain, both because of latitude and longitude and because of approach to great masses of land, also because of the altitude of dominant terrestrial elevation; but it does not follow the rule of any meridian . . . Hence the bounds of variation are not properly defined by great meridian circles."[5]

Observational evidence employed to support this hypothesis (if used at all) will most likely have stemmed from the North Atlantic Ocean (see fig. 3.1). Northeasting in European waters, northwesting in American quarters, and much debate over the actual position of the agonic meridian dividing the two regions would appear to confirm the notion of continental attraction. By contrast, previously encountered zero declination in South Africa, East Asia, and Middle America would have posed a serious counterargument, as would reigning northeasting along the South American east coast. One advantage it did have over earlier representations was the implicit acceptance of curved isogonics, due to crustal irregularities. It is therefore not surprising that Gilbert flatly rejected Stevin's conjectured sextupole, while praising his haven-finding art.

The first five books of *De Magnete* would have easily sufficed to secure Gilbert's place in history, as far as geomagnetism is concerned. But at the time of publication, it was the contents of the sixth book, on "revolution," that had the most impact. In this part, the author unfolded his cosmological perspective, which was as effused with magnetic effluvia as his own laboratory, and thereby introduced magnetism into the debate between geo- and heliocentrism.[6] As outlined earlier, classical cosmology worked from the assumption of a fixed Earth at the center of the universe, inside nine spherical shells, which carried the known members of the solar system and the "Prime Mover," which kept the system in motion. In order to account for the planets' apparent intermittent retrograde movements when observed from Earth, scholars additionally assigned them secondary orbits, so-called epicycles, superimposed on their annual paths. In 1543, Polish astronomer Nicolaus Copernicus had challenged this view by presenting a less complex alternative, which placed the Sun at the center, made the Earth one of its planets, and kept the stellar sphere stationary. Both systems featured circular orbits and uniform cyclical movements, but the heliocentrist interpretation dispensed with a number of planetary epicycles.[7]

Fifty-seven years later, Gilbert was quite rigorous in his own deconstruction of the geocentrists' framework. Regarding the stellar sphere, he first presented the familiar argument that a small planet revolving around its axis would require far less energy than the whole universe daily whirling around it in the opposite direction. Next, in order for the "Prime Mover" to be capable of performing this feat, it would have to be made of some material substance to transfer the necessary force, and a small solid sphere seemed less absurd than a huge one. He then went on to reject all celestial shells outright, replacing all heavenly bodies' poles with magnetic ones (axial dipoles). He thus attributed to magnetism the power to maintain both the Earth's constant orientation and its daily rotation. The orderly behavior of other luminaries floating weightlessly in space he similarly deemed to be governed by the constant interaction of their magnetic effluvia. The philosophy was strengthened by the belief that each star, planet, and satellite had its own magnetic Form, and that similarity in this Form between a pair of bodies caused one to orbit the other. The Moon was therefore magnetically bound to our planet, their interacting magnetic orbs dictating the former's path, and the same side constantly facing Earthward. Lunar magnetic forces were furthermore assumed to be responsible for the tides.[8]

This rationale led to a peculiar conclusion, namely that a terrella, having the same magnetic Form as the Earth, would, when freely suspended, not only orientate itself in the meridional plane, but also mimic its planetary mother in performing a daily rotation. The idea is strongly reminiscent of Peregrinus's supposed sympathy of a magnet with the revolving celestial sphere. The proposition was submitted again in 1641 as a chronometer solution to the longitude problem, to the Dutch governor-general in Batavia, Antonie van Diemen. Its proponent, a certain George Konigh, even claimed to have seen a working prototype of such a magnetic clock at a German astronomer's, a "Dr. Johan Stocken of Franckenhausen in Daeringen." Others were, however, less credulous.

Gilbert thus introduced magnetism into the astronomical debate. Although he supported heliocentrism, he was no true Copernican, or even an auxiliary to their cause. His magnetic proof of daily and annual rotation was rather obscure, and heavily mixed with classical beliefs. His rotating terrella concept, moreover, provided an easy target for Jesuit geocentrists in subsequent decades, who were to use its falsification as a proof of the Earth's immobility in space (see below). Gilbert seems to have been in the opposite camp primarily because heliocentrism fitted his own constructs of magnetic motion, not because he was promoting acceptance of the Copernican interpretation by others.[9]

This did not prevent heliocentrists from employing geomagnetism to further their aims. Most notable of these was Graz professor of astronomy Johannes Kepler. An exchange of letters between him and the Bavarian chancellor von Hohenburg clearly demonstrate Gilbert's influence. During the 1590s, Kepler had developed an interest in the Earth's magnetic field; he had read the available monographs on the subject, and had started collecting observations himself as well. After having encountered several depictions of Mercator's magnetic mountain, he attempted at least twice to calculate a cross-bearing of the tilted dipole. However, his hopes of a geomagnetic longitude solution were dashed upon reading Gilbert's work, which convinced him that the dipole was axial, but its field distorted by sizeable elevated continents. The correlation of magnetism with gravity remained puzzling to him. Around 1603 he chose to follow Gilbert in assigning all heavenly bodies a magnetic attraction and in interpreting their magnetic interactions as determining celestial mechanics. But whereas the Englishman's philosophy had been based on "vital energy" producing qualitative changes, Kepler pursued a more productive line of inquiry by assuming quantified mechanical energy effecting quantitative changes. His magnetism was thus non-vital, and, once reformulated in mathematical terms, led to his famous three laws of elliptical planetary motion in 1609. In addition, he explained only annual motion by magnetic attraction, while considering the Earth's daily rotation a nonmagnetic effect.

Another proponent of heliocentrism was Galileo Galilei, who discussed the subject in his 1632 *Dialogo sopra i Due Massimi Sistemi del Mondo Tolemaico e Copernicano.* Although the Italian scholar regretted *De Magnete*'s lack of mathematical foundations, he nevertheless adopted magnetic force as maintaining the constancy of the Earth's axial tilt relative to the ecliptic. As Gilbert had earlier, Galilei then went on to stress the distinction between crust and core, assuming the Earth's deep interior to consist of lodestone under extreme pressure.[10]

In English navigation, many experts promulgated Gilbert's notion of continental needle attraction. On the European mainland, his intellectual heritage influenced natural philosophers such as Christiaan Huygens, Marin Mersenne, and René Descartes, as well as writers on navigation like Guillaume Denys in France and Jan Verqualje in the Dutch Republic. Gilbert's causal hypothesis of irregular magnetic declination due to the distribution of continental crust was in its day probably the most widespread of all static field concepts. Ironically, his planetary axial dipole with crustal deviation was primarily derived from terrella experiments and deduction, rather than through induction from measurements of the Earth's field proper. Yet its power to accommodate "unexpected" needle readings was far greater than any scheme featuring meridional isogonics, due to, as Thomas Browne (1646) phrased it, "the inequalitie of the Earth variously disposed, and differently intermixed with the sea: withall the different disposure of its magneticall vigor in the eminencies and stronger parts thereof; for the needle naturally endeavours to conforme unto the meridian, but being distracted driveth that way where the greater and most powerfuller part of the Earth is placed."[11]

Notwithstanding the appeal of Gilbert's philosophy, many seventeenth-century scholars failed to be smitten by it. Francis Bacon (who opposed Copernican doctrine) criticized Gilbert's attempt to construct a complete metaphysical cosmology based on a single phenomenon. Actual measurements at odds with Gilbert's proposed continental attraction were, moreover, put forward by seafarers William Baffin and Laurens Reael, from observations made in Baffin Bay (1616) and China (1651) respectively.[12] The most sustained attack was, however, launched by several members of the religious society founded by St. Ignatius of Loyola, defenders of the faith in a geocentric universe.

The Jesuit Response

The Roman Catholic order of the Society of Jesus was founded in 1534 and won papal approval six years later. In addition to missionary, educational, and charitable works, it also played an important part in the Counter Reformation. From a modest membership of about one thousand individuals around the mid–sixteenth century, it managed to grow fifteenfold within a hundred years, and eventually became the largest and most preeminent of all male religious orders in Europe. To join, a clergyman had to take a fourth vow to defend the pope, in addition to the regular three of poverty, chastity, and obedience. The promotion of the faith overseas soon created a network that spanned all corners of the globe. In Europe, Jesuits acted as preachers and catechists among the populace, but also served as confessors to royalty. However, their zeal in championing the pontiff, their broad spectrum of activities pervading many aspects of life, and their growing political power engendered controversy and opposition, eventually causing their suppression by papal brief during the period 1773–1814.

The Society was not only very wealthy, but also intellectually influential. In Catholic countries, the Church largely controlled education in schools and universities through teaching congregations, of which the Jesuits were the most prominent. Although pruned of the worst medieval formalism, the curriculum remained primarily Scholasticist, defending scriptural authority by taking recourse to metaphysical arguments of final causes. When heliocentrism was recognized as theologically heretical, several natural philosophers in the Jesuit order rose to defend papal authority and its geocentrist cosmology. Their efforts included both short, polemic discourses and larger works of encyclopedic scope. They jointly asserted that magnetism was a physical phe-

nomenon that kept the globe fixed in proper alignment at the center of a revolving universe.[13]

Ferrara philosopher, mathematician, and teacher Nicolo Cabeo was the first
among Jesuit scholars to take up the challenge. In 1629 appeared his *Philosophia Magnetica*, a topical treatise endorsing proof by experiments. Since the
Aristotelian interpretation of the world forbade action at a distance, Cabeo,
rather than depending on corpuscular effluvia to transfer magnetic force, postulated rays of magnetic energy instead. These impelled the ambient air to gy-
rate, thereby moving an attracted object. Experiments with a terrella and iron
filings suggested to him that the paths of the emanations simply swept outward
from the poles, instead of following circulatory curves between them. Concerning the Earth, he vehemently opposed Gilbert's animism; he also strongly
disagreed with the Englishman's assertion that the planet was a large magnet
due to a large core of pure lodestone, surrounded by a crust of mixed composition. The Italian argued that the geomagnetic field was too feeble for this hypothesis to be plausible, and surface magnetic declination would similarly have
been negligible if this really were the case. Assuming instead that a magnet received its power from the heavens, he ascribed the strongest magnetic properties to the Earth's surface (centered at the poles), weakening with depth, away
from their revolving source. Cabeo did adhere to Gilbert's interpretation of irregular global patterns of magnetic declination due to the uneven distribution
of large landmasses over the world. This also prevented the determination of
longitude by magnetic means. Cabeo furthermore assigned to the gravitational
force the task of holding the Earth at the center of the universe, whereas magnetism was to restore it to that position after any dislodgement. Magnetic
virtue in other heavenly bodies he did not touch upon.[14]

Polymath Athanasius Kircher was next in the line of major Jesuit adversaries
to magnetic heliocentrist cosmology. A prodigious author and professor of
mathematics at the *Collegio Romano,* Kircher produced a massive tome on
magnetism in 1639, published two years later. The work followed Cabeo in rejecting Gilbert's lodestone nucleus on account of the discrepancy between the
observed and expected ratio of mass versus magnetic force. Kircher likewise
did acknowledge that the Earth possessed magnetic properties, in keeping with
his more general theory that all matter was capable of attraction and repulsion.

According to Kircher, the terrestrial and celestial poles expressed a mutual
affinity in their alignment, ensuring eternal maintenance of orientation.
Gilbert's terrella mimesis he considered "absurd and intolerable," and he like-

wise refused to lend credence to the idea that a spinning spherical magnet would force another in an orbit around it. Since Gilbert had never explicitly placed the Sun at the center of the universe, the Jesuit scholar directed further criticism at Kepler's views, stating that the Sun's magnetic effluvium was inadequate to explain planetary mechanics. For the underlying cause of geomagnetism, Kircher turned to the quasi-organic concept of magnetic *fibers*, transporting magnetic force from pole to pole in analogy to sap inside a plant. These veins followed capricious paths "growing" through the Earth, subject to alteration and destruction by earthquakes, metallic "humours," internal heat, and chemical action by salts. These dynamic subterranean processes not only accounted for perceived irregularities in the disposition of the surface field, but also offered a causal explanation for recently discovered secular variation (published in 1635; see below). Gilbert's supposed continental deviation he considered an erroneous assumption, based both on the limited range of crustal deviation and a number of magnetic observations. A positive correlation of declination with latitude the author similarly discarded as fallacious, following a comparison of measurements made near the Maldives, at Madagascar, and in northern Scandinavia. Kircher was actually in an excellent position to gather empirical evidence, placed as he was at the nerve center of the Jesuit missionary network. Of the fifty-six acknowledged sources of such intelligence listed in the third edition of his *Magnes* (1654), no fewer than forty were members of his own order, some stationed as far away as India and China.[15]

Geocentrism received a further boost through the work of Jacques Grandamy, philosopher and theologian at several French colleges. This Jesuit scholar wrote a *New Demonstration of the Earth's Immobility Begotten from the Magnetic Virtue* in 1645. The content of this fairly succinct tract can be summarized by his central syllogism: "No body having magnetic virtue turns about the poles; the Earth has magnetic virtue; therefore, it does not turn about the poles."[16] It was based on an experimental setup of a celestially aligned rotating terrella, which mainly proved that suspended spherical lodestones make very poor timepieces. It was, however, successful in falsifying Gilbert's claim that the world's daily rotation was a magnetic effect. Amid many rather dismally drawn cherubs, Grandamy then boldly went on to interpret this result as the ultimate proof of a stationary Earth in the cosmos. The author followed Cabeo in designating geomagnetism an orientating force only, while gravitation was supposed to keep the planet in the center of the (nonmagnetic) heavens. In oblique censure of Kepler, he furthermore classified all celestial bodies as lacking any magnetic quality. He also pressed divine arguments into service, recognizing the wisdom of the creator in rendering the globe magnetic, keeping it at a proper stance and distance to receive light and other vital influxes, and thereby maintaining a unique habitat for his creatures.[17]

As a bonus, the postulated axial dipole offered a means to determine longitude at sea. Despite rejecting any regularity in the observable pattern of magnetic declination itself, a combination of locally measured D and I, potentially reinforced by observed latitude, could, according to Grandamy, yield a combination that would be easily associated with a known position. He therefore advocated an official policy by rulers to amass such readings, especially at ports and landmarks. He specifically targeted his own Jesuit order as eminently suited to perform this office, encouraging its members to take note of the phenomenon everywhere. He denied the existence of secular variation, deeming it the result of observational error.[18]

The Jesuits shared with their heliocentrist opponents the acceptance of geomagnetism as a natural force, but despite a growing body of experimental evidence, no consensus followed regarding the structure of the universe. Instead, new priorities beckoned for attention; in a slightly later text by Vincent Leotaud (1668), the Jesuit focus had shifted to do battle with the emerged enemy of Cartesianism (see below). As far as geomagnetism was concerned, Jesuit writers on practical navigation, such as Gianbattista Riccioli (1672), Georges Fournier (1676), Claude François Millet Dechales (1677), and later Yves Valois (1735), stressed the field's irregular characteristics, and the impossibility of thereby determining position on the open ocean. The confusion regarding the causes of the anomalies from a dipole system was particularly apparent in Riccioli's assessment, which blamed a host of factors including the continental crust, magnetic fibers, mines of lodestone and other metals, nearby mountains, subterranean heat, chemical processes, and the Earth's heterogeneous constitution in general.[19] Over a century of data gathering and diffusion, rather than shedding light on the workings of terrestrial magnetism, seemed only to have made matters worse, not least because the whims of the field had by that time been discovered to extend into the temporal, as well as the spatial domain.

"A sensible diminution"

The discovery of a fourth dimension in the variability of the geomagnetic field was a momentous event in the history of science. The subsequent implementation of time dependence in field representations constitutes the most profound watershed in descriptive and causal hypotheses in the analyzed period. Where and when the notion was accepted, a whole class of earlier ideas became redundant at a single stroke. At the same time, new questions related to possible patterns of change beckoned to be addressed. Were local waxing and waning of declination erratic, linear, periodic, or subject to global rotation relative to the crust? What mechanisms actually produced these effects? Could the planet's deep interior still be considered stable at all? And did differences perhaps exist in secular acceleration, in space and over time? What clues about future developments lay hidden in earlier-compiled time series of observations?

Two acquaintances of John Dee were the first to register a local northeasting at London of about one compass point ($11\frac{1}{4}°$). They were instrument maker and teacher of mathematics Thomas Digges in 1571, and five years later Christopher Hall, sailing master for the explorer Martin Frobisher. The exact location of Digges's reading is unknown; that of Hall took place at Gravesend. On 16 October 1580, merchant seaman and later naval commander William Borough was more thorough in collecting and averaging a series of eight observations at Limehouse, which exceeded the earlier-established value by only a few minutes.[20] There was thus as of yet no reason to suspect that alterations were afoot. Around the turn of the seventeenth century, both Wright (1599) and Gilbert (1600) perpetuated this belief; their reiterated value of $11\frac{1}{4}°$ NE may very well have stemmed directly from Borough's text, rather than from their own measurements.

The next stage of the proceedings involved two successive professors of astronomy at Gresham College in London: Edmund Gunter and Henry Gellibrand. Early in 1622, Gunter decided to measure declination at Deptford (some two miles distant from Limehouse), and was startled to find a mere $6°13'$ northeasting. Being aware of Borough's much higher figure, he consulted instrument maker John Marr and other practitioners, and resolved to return to the exact location of Borough's initial measurement to repeat the experiment. Accordingly, on the night of 13 June 1622, Gunter, Marr, and some unnamed companions took two long magnetized needles to Limehouse, where a total of eight readings resulted in a mean of little over $5°56'$ NE. The ensuing year,

Gunter published these data in his work on the sector and the cross-staff. But even though he had thus obtained strong indications of inconstancy, he could not be absolutely certain that Borough, forty-two years earlier, with possibly poorer instruments, had not made a mistake. He therefore abstained from drawing conclusions at that time.[21]

In 1624 a large sundial was erected in the royal gardens of Whitehall, mounted on a cubical stone oriented in the meridian. Two years later Gunter died, and was succeeded by a young mathematician. This Henry Gellibrand initially failed to pursue matters further, and in a short tract on navigation from 1633 rejected the assumption that the magnetic needle could be used to find longitude.[22] It was John Marr who set the ball rolling again that same year. Taking one of Gunter's needles, he went to Whitehall Garden, applied it to the fixed support of "His Majesties Diall," and again found smaller northeasting than on the previous occasion. Being notified of this, Gellibrand assembled a party including Marr on 12 June 1633, and once more brought Gunter's needle to Deptford, in order to be able to compare the present situation with that of eleven years earlier. Five measurements in the morning and another six in the afternoon yielded an average of less than 4°05′ northeasting. On 4 July 1634, Gellibrand confirmed the measurement at St. Paul's Cray in Kent, some twelve miles southeast of London. This time, the magnetic apparatus included two large quadrants and two twelve-inch needles, in addition to Gunter's instrument. Thirteen observations then rendered a combined estimate of about 4°01′ NE. If any doubts had been lingering at the back of Gellibrand's mind, this latest exercise was enough to convince him of the existence of secular variation. It also served to vindicate Borough's earlier measurement as at least plausible. In 1635 appeared the astronomer's treatise *A Discourse Mathematical on the Variation of the Magneticall Needle,* in which he unequivocally stated: "Hence therefore we may conclude that for the space of 54 yeares . . . there hath beene a sensible diminution of 7 degrees and better."[23] Table 3.1 gives an overview of the relevant measurements.

The adoption by others of the new guiding principle for understanding the Earth's magnetic field was neither instantaneous nor uniform. Quite a few longitude-finders continued to postulate static hypotheses after 1635, even excluding the Jesuits (who had cosmological reasons for holding on to time invariance). Some Dutch seventeenth-century navigation manuals likewise persisted in exhibiting skepticism, or presented tables of compiled declinations without mentioning that the figures were liable to change.[24] This is not to say that

Table 3.1 *Observations of Magnetic Declination in London (1571–1634), Leading Up to the Discovery of Secular Variation*

Date	Observer(s)	Location	Northeasting
1571	Digges	London	11°15′
(12 June) 1576	Hall	Gravesend	11°30′
(16 Oct.) 1580	Borough	Limehouse	11°18′53″
1622	Gunter	Deptford	6°13′
(13 June) 1622	Gunter, Marr	Limehouse	5°56′38″
1633	Marr	Whitehall	"less"
(12 June) 1633	Gellibrand, Marr	Deptford	4°04′49″
(4 July) 1634	Gellibrand	St. Paul's Cray	4°01′23″

scholars and navigators in the Dutch Republic remained unaware of the finding. The very first static geomagnetic concept submitted as a longitude solution to Dutch officials after 1635 was explicitly rejected because secular variation had not been accounted for. A later work on navigation by Hellingwerf (1694) seems of two minds, copying some values from the Plancius compilation (then about a century old), while at the same time stating that the field's changeability prevented the formation of a definitive table, and warning masters to trust only direct observation to correct their dead-reckoning. Meanwhile, a new realization was emerging, that instead of such lists of observations being useless, data needed to be collected with double vigor, in order to record field mutations and to arrive at an understanding of the processes at work. This conviction may very well have originated in navigational practice, wherein the need to keep track of changing declination was also recognized.[25]

The discovery of secular variation had been made under controlled conditions by a highly esteemed English scholar. Acceptance by his London colleagues, mathematical practitioners, and other compatriots was consequently far more rapid than in the Dutch Republic. In the very year of Gellibrand's publication, mathematician John Pell drew up a manuscript (*Exercitatio de Diminutate Variationis Causa*), which he circulated among acquaintances to profess his adherence to the new creed. Two years later, Richard Norwood added a reference to secular variation in his *Sea-man's Practice* (1637). As subsequent measurements in London and elsewhere continued to bestow additional confirmation, and no geocentrist requirements forced adherence to a static axial dipole system, there was little obstruction to the idea's being spread and gaining a firm foothold within a few years.[26]

The transition went a little less smoothly in France. After Pell had communicated Gellibrand's discourse to scholar Marin Mersenne in 1640, a French translation appeared, which reached not only Jesuit colleges but also non-Jesuit philosophers such as René Descartes, Pierre Gassendi, and Pierre Petit. In contrast to Grandamy, who contested Gellibrand's observational competence on cosmological and metaphysical grounds, members of this second group initially questioned the conclusion—while appreciating its potential—on the basis of possible deviation due to nearby iron, and differences in magnetization and lodestones, primarily with respect to Borough's value, which was over half a century old and had been obtained under less than adequately detailed circumstances. But eventually these points lost their significance, as the weight of new datasets and the resolution of discrepancies with older ones neatly fitted Gellibrand's claim. Using French data, Gassendi was able to replicate the former's conclusion in 1640, and Petit, upon examining earlier compilations, eventually followed suit, reinterpreting small differences as signal where earlier he had discarded them as noise. In the French navigational literature, the concept was established by the 1660s at the latest. At that time, Denys cited it in his navigational tract as the sole reason for not including a table of magnetic declinations of questionable age.[27]

Time-dependent change not only warranted a reappraisal of seemingly erratic global distribution patterns, but additionally opened up a new avenue of research, namely the investigation of potential regular change over time. Numerous individuals spent prodigious amounts of time and energy sifting through the ever-growing paper legacy of past experience to build new hypotheses, or to find corroborating evidence for a previously devised construct. Among them were many longitude-finders, most of whom assumed one or more magnetic dipoles to perform some kind of subterranean precession around the Earth's rotation axis. Other investigators, however, remained more doubtful, suspecting some regulatory principle to be at work on a global or continental scale, but failing to cast their suspicions into a causal form. From the various time series of unevenly scattered points, they attempted to discern and predict constant rates of change, for instance, related to Atlantic "westward drift," and the changing path and shape of agonics across the globe.

Around the mid–seventeenth century, Dutch scholar Johannes Holwarda stated his belief in no uncertain terms, affirming that declination "changed from time to time with good proportion so that one may justly say that Nature in this respect is found to be very constant in its inconstancy."[28] A letter by Petit to the Royal Society in London from 1667 offered the Frenchman's conjecture of one degree of change for every seven to eight years. In April 1670, the Society also received a report from French natural philosopher Adrian Auzout, who suggested possible proportional change of declination at Rome, while astronomer Hevelius a few months later communicated to London the rate of constant decrease he had established at Danzig (4°15′ in twenty-eight years), and added that future observations would be required to confirm regularity at all times in all places.[29]

The French were particularly zealous in the pursuit of these postulates during the eighteenth century. In hydrographer Guillaume Delisle, devotion reached almost fanatical proportions. Based upon what few of his manuscripts have been preserved, it is possible to get an idea of the task he set himself. Not only did he compile some ten thousand observations by hand (from logbooks and published sources), he also tried, over a period of about twenty years, to analyze and calculate regional rates of change and acceleration for the whole world. He never attained his dream of combining his widely varying figures into a single system, but continued to cling to the hope that more data points would finally lift the veil. The disappearance of the majority of his manuscripts constitutes a profound loss for the study of geomagnetism.[30]

Jesuit scholar Nicolas Sarrabat (1727) instead tried to impose some order in the growing pile of declination readings by making the distinction between general geomagnetic phenomena (long-term, observed nearly everywhere and without interruption) and particular ones (limited to certain times and places). He directed most of his attention to the changing behavior of lines of zero and maximum declination, looking for "a sufficiently constant progression, and a behavior that is regular and uniform, provided that some particular influences do not oppose it."[31] Despite more skeptical opinions in the second half of the eighteenth century, some French authors continued to keep this thought alive. The 1761 *Mémoires de l'Académie Royale des Sciences de Paris* noted a yearly increment of nine minutes. Similarly, in 1775 mathematician Etienne Bézout estimated a fixed annual increase of about ten minutes northwesting for the French capital. However, a few years later, Le Gaigneur, in his *Pilote Instruit* (1781), was a lot less optimistic in its assessment; he reminded his readership that isogonics tended to move with variable speed, that their position at sea was merely based on rough estimates of longitude, and that "in certain places the variation testifies to continual incertitude, which has no constancy whatsoever."[32]

Magnetic Particles

The mechanistic philosophy put forward by René Descartes in 1644 attempted to explain all natural phenomena by reducing them to principles of matter and motion. Like Mersenne, Descartes had been taught at the Jesuit college of La Flèche, and one can interpret the rationalist physics he developed in part as a reaction to scholastic Aristotelianism, which depended on metaphysical arguments to describe the universe. Descartes instead evolved a theory based on rigid, inert, submicroscopic particles (*corpuscules*) of various shapes, constantly gyrating and forming vortices at all scales. These filled the cosmos with a transparent *ether*, making matter and space synonymous in the absence of a vacuum. In this "fullness" of infinitely divisible matter, all motion was passed on through immediate contact, and all energy was conserved, leading to a deterministic interpretation of the material world. Cartesians thus supplanted the search for final causes by an inquiry into the mechanical laws governing the effects of motion. The ubiquitous turbulences were supposedly responsible for all interactions between bodies.

The Cartesian explanation of magnetism was particularly imaginative, involving elongated minuscule magnetic particles twisted like a screw, circulating through and around lodestones. While constantly rotating, the subtle matter supposedly threaded magnetic objects through special conduits, which were aligned with the polar axis and configured to receive the flow in one direction only. Like modern road and rail tunnels, a parallel system of channels accommodated opposing streams, and thereby allowed the establishment of a never-ending double circulation through and around a magnet. Magnetic repulsion was thought to result from collisions of opposing vortices, while attraction purportedly came about by the flow forcing away the air between magnet and drawn object. As regards the magnetization process, Descartes postulated that the internal pores of nonmagnetic iron would branch in all directions, requiring the vigorous stream from a nearby magnet to align them.[33]

The revolving ethereal currents he moreover imagined instrumental in driving celestial mechanics. A solar vortex swept the planets along in a whirling spiral of magnetic matter. Smaller swirling systems prevented moons from straying into deep space, and transferred tidal energy. Descartes furthermore followed Gilbert in assuming the whole Earth to be a magnet, with maximum strength concentrated between the poles, and its magnetically receptive pores parallel to the rotation axis. The relative weakness of the surface field the philosopher subsequently explained by the idea that many corpuscles found suitable routes at depth, worming their way through underground layers of metallic ores and other easily adaptable mantle materials, without ever reaching the atmosphere. For declination, Gilbert's crustal inegalities once more came to the rescue. The cause of time-dependent change was sought in the generation and deterioration of iron mines.[34]

Seventeenth-century France to some extent mirrored classical Greece in counting among its inhabitants the foremost proponents of both effluvial and atomistic mechanistic philosophy. As Descartes continued the line of reasoning initiated by Empedocles, Plato, and Plutarch, so did atomism, as put forward by Democritus, Epicurus, and Lucretius, find a worthy early-modern representative in Pierre Gassendi. Philosopher, theologian, astronomer, and mathematician, Gassendi also experimented for years with magnets. In his philosophical writings, he stressed the utility of both inductive and deductive analysis, and extensively criticized Cartesian doctrine. The presented alternative was equally mechanistic, but used changing aggregates of immutable, indivisible particles, rather than vortices of infinitely divisible corpuscles, to interpret nature's workings. He listed magnetism's variable strength, its decay with time, and its annihilation by heat as proofs of corpuscularity. The explanation of (geo)magnetism did not significantly differ from the one employed by Descartes.[35]

The Jesuits formed a second group of opponents to Cartesianism. Their members raised many objections and injunctions against the philosophy, which they judged subversive on account of its methodical doubting. Some Cartesian tracts even ended up on the Papal Index of Forbidden Literature. All was to no avail, as the new insights continued to gain ground, while themselves facing new competition from emerging Newtonian philosophy in later years. Jesuit authors reluctantly trailed these developments, and came to slowly assimilate corpuscularism over time. None of these representations entailed enough regularity in global patterns to determine position at sea.[36]

French ideas concerning magnetic particles enjoyed a more favorable reception in England. Their initial communication across the Channel is largely due to Kenelm Digby, a man of many trades. Naval commander, diplomat, adventurer, and amateur natural philosopher, he also became an English exile in the 1640s. In Paris he witnessed a French scientific revival, being present at gatherings in Mersenne's cell in the Convent of the Annunciation and the private lodgings of such learned individuals as Henri Louis Habert de Montmor and Melchisédech Thévenot. Among the invited guests were Descartes, Gassendi, Thomas Hobbes, and Blaise Pascal. Using what he had picked up during their discussions, Digby compiled two philosophical treatises which were eventually communicated to a wider English audience. Of these, *The Nature of Bodies* (1644) contained an atomistic theory of magnetism. Digby postulated a circulation of magnetic particles around and through the Earth, driven by atmospheric dynamics. Due to the Sun's heat, rarefied air in the tropics rose to flow polewards at altitude, while colder and heavier polar air went the opposite direction near ground level. This conveyor belt was thought to carry magnetic matter across the world. Conglomerates would form lodestone concretions at the surface. Some particle "rivolets" alternatively went into the Earth, fostering a Gilbertian core magnet. In addition to atmospheric influences causing irregular declination patterns, crustal heterogeneities were likewise to blame. An interesting detail was Digby's comparison of magnetic orientation to a weather cock butting its nose into the wind, incorporating the concept of local flow, rather than distant poles, guiding the needle.[37]

In 1666 the informal association of Paris researchers that had earlier inspired Digby received a more permanent footing; upon the suggestion of administrator Charles Perrault to Louis XIV's minister Colbert, the Académie Royale des Sciences de Paris was founded. Their state-funded members met twice weekly in the Parisian Royal Library to pursue investigations of potential benefit to the country (including magnetism and navigation). In England, the first semi-official venue for a mixed congregation of scholars, philosophers, and mathematical practitioners was London's Gresham College, already founded in 1597. At the time, Oxford and Cambridge were the only cities with universities, and their curriculum did not yet include many of the practical scientific subjects on offer at Gresham. The London college was particularly successful as a central exchange of knowledge and research in the first decades of the seventeenth century.[38]

By the time the English Civil War (1642–48) erupted, an "Invisible College" had formed, which met to discuss all topics except religion and politics. Upon the ascent of Charles II, a formal scientific society for the advancement of experimental philosophy came into being, which gained a royal charter in 1662. The "Royal Society of London for improving natural knowledge" revolutionized the expression of scientific thought by undertaking publication (from 1665) of the first-ever scientific periodical, the *Philosophical Transactions of the Royal Society of London*. Its wide circulation had a substantial impact on the diffusion, in England and abroad, of data, experimental results, and theory, no longer accumulated and gestated in massive tomes after decades of compilation and private contemplation. Academies on the European mainland were to follow suit later.[39]

It was one of the Royal Society's co-founders, Robert Boyle, who promoted corpuscularism in England throughout most of the second half of the seventeenth century. Together with Henry More, Thomas Brown, Henry Power, Robert Hooke, and Thomas Henshaw, he endorsed a mechanistic philosophy to explain magnetic effects observed in the laboratory. His invention of the air pump enabled him to confirm magnetic action in a vacuum. Iron and lodestones submitted to percussion and extreme changes in temperature furthermore strengthened the belief that a variation in magnetic properties was due to an alteration in the iron's texture, as a result of mechanical operations. The underlying theory bore both Cartesian and Gassendian traits, supporting the idea of a corporeal, atomic effluvium in constant circulation.[40]

A number of reasons spring to mind for why this concept took hold in England at this time. First, the assumption of direct interaction through particle contact seemed more productive than occult Aristotelian physics actuating Form at a distance through a Magnetic Quality. Second, proof of particle flux would refute Gilbertian animism and other vitalist claims at the heart of early magnetic philosophies. Third, such a proof would serve a wider purpose of supplying evidence of corpuscularity in general, embracing much of physics and chemistry. Fourth, Jesuit assertions of incorporeal action linked to a geocentric cosmology provoked a strong response from Puritan heliocentrists in England. Fifth, mechanistic philosophy was very much a child of its time, arising in an age of experimental research programs focusing on motion and change, and novel machinery, automatons, and instruments, which served humankind both in a physical sense and in providing analogies. Lastly, the lively intellectual debate across the Channel provided a welter of tracts, arguments, and opinions, which stimulated experiment and discussion in England.[41]

Several initiatives unfolded from the 1650s to the 1680s to get at the heart of the magnetic matter. The Royal Society's "Magnetics Committee" undertook a series of annual measurements of declination, and performed numerous lodestone experiments in house. Henry Power's treatises (1654–64) outlined a comprehensive investigative program, which attempted to assign responsibility for the observed phenomena to the effluvial medium rather than to the magnet itself. Robert Hooke studied the field's unmitigated force through interposing bodies, and entered into a philosophical dispute with Martin Lister, one of the last defenders of traditional magnetic philosophy. Moreover, in 1684 the Oxford Philosophical Society came into being, providing a new drive to conduct relevant research under magnetician John Ballard. Despite a persistent failure to obtain direct visual confirmation of the existence of magnetic corpuscles, Gilbertian magnetic philosophy was eventually generally rejected, in favor of subsuming hypotheses of magnetic action in a broader mechanistic framework of effluvial dynamics. Magnetism was consequently no longer considered a unique property delineating a special class of materials, and thereby lost its coherence as a separate discipline of inquiry. For the terrestrial field, this spelled a new era of uncertainty, with internal, crustal, and atmospheric disturbances vying for dominance.[42]

The confusion is perhaps best exemplified by the opinion of Isaac Newton on the issue. His monumental work on dynamics and the mathematical formulation of universal gravitation enabled the quantified description of celestial mechanics in terms of gravity, supplanting magnetic cosmologies. If he had applied the same rigor to analyzing magnetic experiments, he might have established the law of mutual magnetic attraction a century before Charles-Augustin de Coulomb eventually did in 1785. Instead, Newton's interest in magnetism seems to have largely originated in alchemical pursuits around the 1670s. Whether he actually adhered to occult principles of sympathy is unknown. Although his theory of gravitation assumes the reality of action at a distance, his work on optics relied in part on the existence of effluvia. The scant references to magnetism in his *Philosophiae Naturalis Principia Mathematica* (1687) are ambivalent, mainly appearing in the context of discussion of other forces. Sometimes it was supposed to act as a force itself, at other times as the result of the mechanical action of subtle matter. Newton's consideration of the geomagnetic field hardly bears referring to as a serious postulate, in that it merely asserts it to be "very small and unknown."[43]

One of the main reasons for studying magnetism was the advancement of navigation; it was one of the primary objectives of the Royal Society, not least in relation to the longitude problem. Fellows William Molyneux, Christopher Wren, Robert Hooke, Peter Perkins, and Edmond Halley were all either directly or indirectly involved in the quest for a solution by means of the terrestrial field. In the very first *Philosophical Transactions* (1665), an article encouraged the seafaring community to supply several types of observations compiled at sea, to improve sailing directions and existing datasets. Shortly thereafter, com-

pass maker John Seller was invited to communicate some of his lodestone experiments to the Society, which, like the magnetic readings contained in ships'
logbooks, provided food for thought for theoreticians.[44] A few of the most
noteworthy geomagnetic hypotheses engendered in this environment will
presently pass review.

The First Dynamic Dipoles

The notion of secular variation initiated the third phase of geomagnetic
thought, and was rapidly adopted in England. Remarkably, for little under a
century English proponents remained the sole purveyors of dynamic magnetic
longitude schemes; their counterparts on the mainland only started to catch
up in the 1730s. Continental scholars did accept the reality of time-dependent
geomagnetic change, but differed from their English colleagues in rejecting the
regularity imposed by one or two dipoles precessing around the Earth's axis. It
is debatable which side of the Channel was more advanced in the second half
of the seventeenth century. Arguably, dynamic systems more closely represented the actual characteristics of the Earth's field than lingering static concepts, but the same can be said for Cartesian vortices when compared to the
simpler view of a single dipole pursuing a circular path. It is a matter of opinion as to which is considered to agree better with reality: the quantitative-
reductionist view or the qualitative-complex approach.

The first Englishman to develop the idea of a precessing dipole (phase
three) was Henry Bond Senior, a mathematical practitioner in the city of London. Although he shielded particulars from public scrutiny for decades, and
others eventually preceded him in full or partial disclosure of similar postulates, his correct prediction in 1639 of zero declination at London in 1657 placed
him far ahead of the competition. His eventual publication of *The Longitude
Found* (1676) was somewhat anticlimactic though, as by that time the system
had already been investigated and found wanting by a royal committee, and
mostly elicited more unfavorable comments.

Bond, who had been a teacher of navigation since 1623, had taken over the
editing of a nautical ephemerides upon the death of its first editor, John Tapp.
It was in the 1639 edition of this *Sea-mans Kalender* that he first intimated his
discovery of "a manifest way (which cannot be stopped but will come to perfection) to attaine the longitudes."[45] He went on to assert the existence of sub-
lunary magnetic poles, of which he claimed knowledge of position and annual

motion, and consequently their period of precession, spanning hundreds of years. After a plea for observations from the western hemisphere, the text ended with the prediction: "Those that live untill the yeare of our Lord 1657 shall not see any variation at all at London and afterwards it will increase westerly at least for 50 yeares."[46] The word "variation" here referred (as in common nautical usage) to magnetic declination.

As years rolled by, the prediction remained unchanged in new editions of the *Kalender*. In 1648, a letter from Charles Cavendish brought Bond to the attention of John Pell, describing the practitioner as old and humble, but an able mathematician nonetheless. The following year, the modest man in question solicited the support of several scholars, who signed his petition to the Dutch East India Company. After introducing himself and his longitude solution in it, he requested employment on board one of the company's eastbound vessels to verify his hypothesis. He suggested a wage of forty pounds a year plus ten for instruments, deferring any major reward until such time as the scheme would come "to bee published for your honor and the publique good of your nation."[47] No trace of this request has been found in Dutch archives, so it is not certain that Bond actually submitted it.

The year 1657 is notable in the present context for two reasons: Bond's prediction of zero declination at London came true, and another mathematical practitioner took over the editorship of the *Kalender*. Shortly thereafter, this Henry Phillippes appended a paragraph to Bond's, affirming the validity of the foretold change in declination sign. Surprisingly, he then presented some of his own ideas: secular variation he estimated to continue for thirty to forty years at about eleven minutes annually, while he set the period of the dipole's orbit at 370 years. It would take until 1676 before Bond publicly distanced himself from this assessment, affirming a period of 600 years instead.[48] Despite Phillippes's added favorable comments on Bond and his scheme, it is difficult to shake off the impression that it was an attempt to take credit from the old practitioner. It would not be the last time.

Meanwhile, the Royal Society had become aware of Bond's prophecy, and in 1662 this body invited him to join its Magnetics Committee for the annual declination measurement at Whitehall. In following years, various Royal Society Fellows engaged in the yearly verification of Bond's predictions, based on declination tables which the author was at that time still unwilling to unveil. Only in 1668 could Pell finally persuade Bond to publish a table of predictions for London up to 1716 in the *Philosophical Transactions*.[49] In hindsight, the slow alteration of local declination—previously a handicap in obtaining proof—had become a benefit, since a significant departure from Bond's predictions would take years to establish.

The pace of events quickened in the 1670s. In 1672 Bond finally committed his draft proposal to paper and supplied Pell with a copy. The year thereafter, the latter was able to convince its author that it was high time to publicize the complete scheme, so that it could be properly examined. Accordingly, an announcement appeared in print, detailing that after thirty-eight years of study the secret of longitude had at long last been discovered, by means of a geomagnetic hypothesis that had accurately predicted the preceding ninety years of secular variation at London. In addition, the text claimed that a table of calculated inclinations had been successfully tested at four widely distant locations. Upon suitable generosity all would be revealed. Pell, who by that time had taken up the role of promoter, submitted several copies of the text to various officials, and was most likely directly involved in gathering support at court for assembling a royal committee of investigation. It eventually formed the following spring, and consisted of William Lord Brouncker, Silius Titus, Seth Ward, Samuel Morland, John Pell, and Robert Hooke.[50]

But before the wheels of state were set in motion, Hooke surprised everyone on 19 March 1674 with the introduction of his own geomagnetic longitude solution. Based on an annual change in declination of between ten and eleven minutes, his internal precessing dipole performed an orbit with a radius of ten degrees in the space of 370 years. Indeed, aside from the added colatitude parameter, the system was identical to that of Phillippes over a decade before. To boot, Hooke somewhat sanctimoniously declared "that he knew not, whether it was coincident with that of Mr Henry Bond."[51] Even if by that time he had not yet received further particulars on Bond's postulate, the fact remains that he had been involved for the previous ten years in its verification, and a person of his mathematical ability must be deemed perfectly capable of connecting the dots on a polar plot, and drawing his own conclusions.[52]

Meanwhile, the committee set down to work in earnest in April 1674, convening and corresponding at several instances over the summer. Seth Ward acted as a liaison between its members and King Charles II, who had become personally interested after a meeting with Ward, Pell, and Bond. In order to gain financial endorsement for instruments and measurements, Ward subsequently attempted to explain the system to the king amid his entourage a number of times, and was eventually granted a private audience. By that time, Bond had finally drawn up an abstract of his theory, which he sent to Ward, together with an annotated excerpt written by Pell. Bond later drafted a second paper to clarify certain matters further.[53]

The reaction of Charles to the proposal is not known. From December 1674, most of the committee's members became involved in the new task of judging a lunar longitude proposal submitted by a Frenchman, the Sieur de St. Pierre. Bond was left to fend for himself, and had by 1676 succeeded in scraping together enough money to fund the publication of his *Longitude Found*. For all his troubles, he was eventually awarded an annual pension of fifty pounds. A presentation copy for the English East India Company earned him another gratuity, which must likewise have contributed to the brightening of his twilight years.[54]

Bond's system consisted of a precessing dipole at colatitude 8°30′, taking 600 years to complete a revolution (see fig. 3.2). It was the first of many variations on a theme. The inventor believed that longitude could be determined using the known position of the magnetic dipole and his tables, listing magnetic colatitude for every five minutes of inclination. A collection of ninety-seven observations at various locations further strengthened this premise. In addition to the reliance on inclination rather than declination, another remarkable aspect of the postulate concerned the magnetic poles, which Bond situated in the sky, forming part of a magnetic sphere that encompassed the Earth. The planet (spinning counterclockwise) supposedly transferred angular momentum to it, but owing to this sphere's "being a substance that hath not solidity to keep pace with the motion of the Earth," it would rotate with less speed, resulting in a slow clockwise revolution relative to the surface. In an interesting reversal of Jesuit arguments, Bond interpreted the revolving nature of the magnetic sphere as strong evidence of the globe's daily turning upon its axis.[55]

Early in 1677, the Royal Society received an anonymous letter criticizing the proposal on a number of grounds. It questioned the supposed longitude of one of the four example locations, as well as the founding of a global system upon a handful of observations, when so many other potential data points existed for which coordinates had been accurately determined. The unknown author then continued by suggesting that Bond lacked objectivity, being "so fond of his owne supposition of things th.t he was unwilling to put his hypothesis to this kind of tryall."[56] This was a mild response compared to the treatment given in Peter Blackborrow's *The Longitude Not Found,* which came out in 1678, after

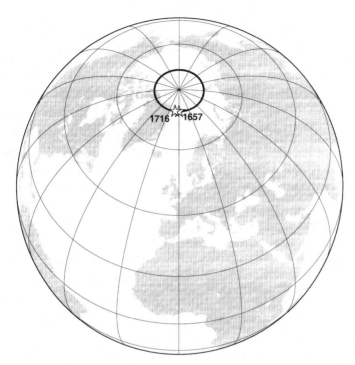

Fig. 3.2. Bond's precessing dipole (1639) revolved clockwise around the geographical pole in 600 years, passing the London meridian in 1657; the postulated position for 1716 is also shown. Orthographic projection.

the old practitioner had expired. This critic considered the whole foundation of the work to be an airy imagination. His most powerful line of attack was a table of ninety-three observed inclinations compiled during the previous two years, which displayed consistent gross discrepancy with the longitudes obtained through Bond's method.

Nevertheless, the first-conceived precessing dipole scheme refused to die. It continued a lingering existence in navigational instruction manuals such as Seller's *Practical Navigation* (1689), and during much of the eighteenth century in Atkinson's *Epitome of the Art of Navigation*. While other parts of this text were updated to reflect innovations, the paragraph on Bond's precessing dipole remained unchallenged in new editions of 1707, 1744, 1753, 1757, and 1767. By this time, the system was 91 years old in its published form, and 128 years when reckoned from its first inception. It was finally put out of its misery by John Adams in the 1790 revision of Atkinson's work.[57]

Moving Multipoles

The final two decades of the seventeenth century formed the cradle for the extension of dynamic concepts to multipole constructs. The development bears some analogy to the late-sixteenth-century introduction of additional pole pairs in static systems; both tried to account for observed field characteristics exceeding dipole complexity by introducing additional dipoles. The intellectual debate concerning terrestrial magnetism during the 1680s in England has been described above as an era of uncertainty, following the collapse of magnetic philosophy as an independent discipline. Whereas some pessimists despairingly deemed noise to dominate signal, others again sought refuge in the multiplication of poles. Unfortunately, in the absence of a quantified law of magnetic force, philosophers could only assess the contribution of each distant concentration of field lines to local declination in a qualitative sense. This did not preclude the formulation of sophisticated hypotheses, but did impede their possible predictive power. Many of the postulated dynamic multipoles therefore failed as practical solutions to the longitude problem. Thus, ideas without explicit navigational aims in mind came to the fore, serving primarily to advance understanding of the planet's inner workings. The suppositions of Peter Perkins (1680) and Edmond Halley (1683 and 1692) belong to this latter category.

Perkins had been a teacher of mathematics in Guildford before becoming the appointed master of the Christchurch Royal Mathematical School in February 1678. This institution was set up to create a small elite of navigational practitioners; a maximum of forty boys received instruction in mathematical navigation for four years from age fourteen, before being apprenticed to masters at sea. Many scholars and writers on navigation had dealings with the school in one capacity or another, among them Astronomer Royal John Flamsteed, Admiralty Secretary Samuel Pepys, and Robert Hooke.[58]

When Perkins took on the assignment, he came highly recommended, both in teaching skills and scholarly prowess. From 1679 he collaborated with his friend Flamsteed and with Halley on the publication of the school's textbook, for which he wrote the section on navigation. They also discussed mathematical problems with each other, and in 1678 Perkins and Halley assisted Flamsteed at Greenwich in the observation of a lunar eclipse. Through the two astronomers, Perkins quickly became part of a wider circle of Royal Society Fellows, who met at London coffeehouses. Once his knowledge of magnetic measurements had spread, the invitation to join the Society came on 5 February 1680.[59]

That same month, Perkins gave a detailed account of declination observations to the Fellows. An animated discussion with Hooke ensued on the effect of inclination on magnetic measurements in the horizontal plane, which was continued when the party reconvened the week after. It was during these exciting weeks that Perkins unfolded his personal geomagnetic hypothesis. In a letter to Richard Towneley dated 13 February 1680, Flamsteed wrote:

> Mr Perkins had collected a vast number of observations of the variation made in seuerall places from printed and manuscript journalls, from which hee tells me hee finds and has traced 4 seuerall meridians on the globe, under which theire will be none; that the needle has no certeine pole th.t may be properly called so to which it allwayes pointes, but in euery different latitude respects a different point in the Earth; that these haue a motion much slower th.n Mr Bond has stated and that no sphaericall triangles will answer the problemes of the variation.[60]

This private announcement thus constituted a four-pole system. Flamsteed's use of the word "meridians" is puzzling, since it conflicts with the statement that spherical trigonometry was unable to furnish a prediction of local declination, and the then available datasets likewise no longer supported the simplifying assumption of meridional agonics. Perhaps the author had agonic curves in mind. Reconstruction of Perkins's concept becomes even more problematic when we take into consideration a remark that the inventor uttered at the aforementioned meeting, on 23 February. Thomas Birch, the official historian of the Royal Society, stated that "he had found by observation that there were six meridians in which the needle did not vary, three in the north, and three in the south; and that one of these went now through St Helena."[61] A division of patterns in the northern and southern hemispheres suggests that the notion of antipodality may have been given up, placing agonics in each half of the globe at different longitudes. The word "now" once more stressed the time-dependent change the field was subject to.

The two conjectures, voiced practically simultaneously by Perkins, leave much to be desired from an analytical perspective. Apart from the fact that they are inconsistent with each other, vital details are missing, such as the actual location of the poles, their path, speed, and period of precession, as well as the imagined global pattern of isogonics. These omissions may be due to the preliminary nature of the postulate(s); perhaps Flamsteed had been premature in persuading Perkins to make his opinions known. The need for further empirical study was evident. In early March the idea arose at the Society to peruse all available journals of sea voyages to extract useful information. Shortly thereafter, Perkins was entrusted with compiling these data, starting with printed accounts. Already on 17 March he produced a preliminary list of sources; disappointingly, most of these had turned out to be scarce. He had similarly found few manuscript logbooks of use at the East India Company, but large quantities of material at the Navy Office. The Royal Society subsequently asked him to continue his inquiries and to provide regular updates on his progress, which he promised to do. Presumably, Perkins proceeded with data processing and analysis throughout the rest of the year. Regrettably, he was never to cast his notions in a more elaborate form, because he suddenly died on 12 December 1680. It was a great shock, especially to his friend Flamsteed. In later years the research notes he bequeathed would have a lasting impact on the relationship between the Astronomer Royal and Halley (see below).[62]

Some two and a half years later (on 23 May 1683), it was Halley's turn to address the Royal Society on geomagnetism. After performing an experiment with a needle and two lodestones, meant to demonstrate that a compass would follow whichever magnetic source was nearest and strongest, he mentally partitioned the globe into four regions. Each area he imagined to be ruled by a separate magnetic pole, placed at specific coordinates (see table 3.2).[63] The last pole was supposedly much stronger than the other three, adding another factor of complexity. The arrangement furthermore ruled out a dissolution into two independent antipodal dipoles, the points of attraction being placed in a decidedly unbalanced fashion. This feature appears to indicate that the idea was data-driven to a considerable degree.

In a paper published in the *Philosophical Transactions* that year, Halley made some substantial additions to flesh out his theory. To start with, he produced a table of fifty-five dated declination observations all over the planet; five of them at London over the years, nine of them at sea, forty-seven locations in all. He then proceeded by rejecting Bond's longitude scheme, dipole representations in general, Gilbertian continental attraction, Kircher's magnetic fibers, and Cartesian mutating bodies of iron ore underground. Instead, each of his four poles he assumed locally dominant, although effects by the others were not necessarily negligible. Lacking sufficient data points for exact geometrical determination of their coordinates, Halley presented the polar positions in his

Table 3.2 Halley's 1683 Hypothesis as Presented in His Royal Society Lecture (23 May 1683)

Colatitude	Meridian	Affected Area
7° N	Land's End, England	Europe, Tartary
16° N	California	Azores, N. America, Japan
[15° N]	[mid-California]	
16° S	20° W of Magellan Str.	S. Atlantic, S. America,
	[95° W of London]	S. Pacific
20° S	Australia and Celebes	Cape of Good Hope,
	[120° E of London]	S. Indian Ocean

Note: Halley's alterations, published in *Phil. Trans.* 13 no.148 (1683): 208–21, are given in square brackets.

paper only as approximations, nearly identical to the ones earlier given in his lecture (see table 3.2: bracketed alterations).[64]

Mathematical resolution of geographical uncertainty would have to await the amassing of a large number of contemporaneous land-based observations, made "with greater care and attention that [*sic*] the generality of saylors apply."[65] Nevertheless, secular variation would, in his opinion, still prevent the establishment of a complete doctrine of the "Magnetical System" for centuries to come. Once more extrapolating from his tabulated declinations, Halley explicitly assigned all four poles a westward motion. The discovery of their true paths he reckoned to be "reserved for the industry of future ages."[66]

Historians of science have remarked upon a decided cooling in the formerly friendly relationship between Flamsteed and Halley, which took place between 1683 and 1686. Now that recent efforts have made accessible Flamsteed's correspondence, more insight is gained into the likely chain of events leading up to and reinforcing the rift between the two. Several factors seem to have brought about a growing animosity, one of which being Flamsteed's allegation concerning the originality of Halley's 1683 hypothesis.[67] The evidence presented here is mostly based on Flamsteed's papers, and may therefore be colored by his personal view and interests. But the witnesses from whom he gathered his testimonies seem to have had no particular reason to bear Halley ill will. The reader is invited to judge.

Around the first of January 1681, about a fortnight after Perkins had been buried, his two brothers Eysum and Thomas paid a visit to his widow at Christ's Hospital, presumably to sort out the affairs of the deceased. Their former sister-in-law then informed them that a few days earlier, Halley had come to peruse her late husband's papers, and that he had offered her seven or eight shillings for the substantial part that held his interest. She also stated that he had been willing to pay her more, but that she had rejected this on the grounds that the proposed sum was quite satisfactory. The rest of Perkins's notes she proffered to Eysum, who took some of them.

Later, Flamsteed too decided to pay the widow a call, with the same intention of securing Perkins's work on geomagnetism. It must have been a disappointment for him to find that Halley had already procured a sizeable portion of the material, softened a little by Mrs. Perkins's graciously allowing him to take the rest. Upon becoming acquainted with Flamsteed's interest, Eysum subsequently donated his brother's writings on geomagnetism to the Astronomer Royal as well. In October of that year, Flamsteed mentioned in a letter to Molyneux "the worke of my deceased friend Mr Perkins . . . saved by me by great accident."[68] At the time, Halley's actions remained unspoken of. Given that his correspondence in later years was to express suspicion, accusation, and reproach of Halley, their absence in this letter is notable, and shows that he possibly did not yet harbor ill feelings toward his colleague at that moment.

The situation had radically worsened by 1686, when Flamsteed wrote to his friend Towneley about Halley: "His discourse in the former transactions concerneing the variation of the needle and the 4 poles it respects I am more then suspicious was got from Mr Perkins . . . who was very busy upon it when hee died . . . Mr Hally was frequently with him and had wrought himselfe into an intimacy with Mr Perkins before his death, and never discourst any thing of his 4 poles till sometime after I found it publisht in the transactions."[69] One has to recall that the astronomer had already notified Towneley of Perkins's four-pole hypothesis on 13 February 1680, so when Halley unexpectedly presented a similar postulate to the Royal Society without acknowledgments, little over a year after he had bought the majority of Perkins's work on the subject, surprise must have easily given way to profound suspicion of duplicity. Questioned by Flamsteed, Halley neither conceded any wrongdoing nor allowed access to his part of Perkins's legacy to prove his innocence. Flamsteed was left to brood over Halley's "art of filching from other people and making his workes their owne" in a private memorandum (1695), not receiving much support from others around him.[70]

The controversy did not end there. Late in 1700, Halley had returned from two Atlantic voyages to study magnetic declination at sea, and his first isogonic chart of that region was nearing completion. It may have been this impending publication, or else the continued acrimonious exchanges between the two scholars, that finally persuaded Flamsteed to undertake some detective work in order to further substantiate his allegations. Accordingly, he wrote to Perkins's brother Thomas, and requested him to search his memory for details of Halley's purchase in December 1680. The astronomer also took the opportunity to ask the brother to convey the same query to Perkins's widow.[71]

Thomas Perkins's reply contained the story conveyed in the previous paragraphs, and held no new information for its recipient, except regarding the amount paid by Halley for the papers. The widow's response, however, added a new twist to the tale. She now blamed Perkins's brother Eysum for delivering to Halley a prime selection of books and manuscripts, and claimed she could no longer remember how much money she had earned in the bargain, but was nevertheless confident that it had been only a fraction of the actual value. Apparently, this result satisfied Flamsteed. In a letter to Wren two years later, he asserted, speaking of Halley: "I have some papers in my hands that prove him guilty of disingenuous practises." But despite this "proof," he seems to have undertaken no further action. Flamsteed continued in his position as Astronomer Royal until his death early in 1720. The fact that his successor was to be none other than the man he had so long suspected of plagiarism must have made him turn in his grave.[72]

With all the actors in this drama long since bereft of life, and Peter Perkins's own manuscripts lost, there seems little chance of a balanced assessment as to whether Flamsteed was right. Modern historians are at odds on it. Taking into account that Perkins's theory was still under construction at the time of his death, Halley may very well have evolved his own multipole theory based on the dataset compiled by Perkins, perhaps inspired by the latter's solutions; by applying similar techniques, he could have reached a similar answer. This would explain where Halley found the empirical support (the table) for his theory, as well as the reason for his avoidance of the question of whether he had copied Perkins's system. His failure in the 1683 paper to acknowledge Perkins as a source of inspiration and data most likely harmed his relationship with Flamsteed, destroying much of his credit in the eyes of the Astronomer Royal. The enmity could possibly have been prevented if he had attempted to redress this situation, even at a later time. Flamsteed for his part can be held responsible for continuing to nurture a grudge for decades, resentment that lost its significance from the perspective of scientific developments when Halley presented his second geomagnetic postulate in 1692, which was definitely his own.

Like the previous attempt, Halley's second effort to understand the Earth's magnetic field had no direct bearing on the longitude problem. In a discourse at the Society on 9 May 1688, he had examined merits and flaws of many potential longitude solutions, concluding that only perfection of the lunar theory held promise. Some three years thereafter, in November 1691, he presented the first outline of the new geomagnetic hypothesis, illustrated with a figure the following week. The next month was probably spent writing up the new ideas, for in January he publicly read the first part of the resulting essay, followed by the second part in May. Once again the *Philosophical Transactions* provided the vehicle to spread his theoretical musings.[73]

The central premise assumed secular change to be "gradual and universal, and the effect of a great and permanent motion."[74] In an interesting aside, Halley rejected the possibility of a magnetic fluid in the core—a concept much closer to the true origin of the internal field—because at the time no liquid magnetic substances were known.[75] The author therefore resorted to a solid nucleus revolving inside an equally solid shell, sharing a common center and spin axis, and separated by a fluid medium. Both bodies carried a dipole, the one in the crust being fixed, the one in the core moving relative to the surface. A slight difference in daily rotation would in the course of several years cause a sensible change in the experienced field everywhere. Presuming this eastward motion to be transferred from the outside inward, the nucleus would lag somewhat behind, producing a commensurate westward shift. The short interval of observed declination, in combination with crustal deviation, prevented determination of the actual speed of movement. A preliminary rough estimate put the period of the internal dipole around 700 years. The author furthermore maintained his belief in the polar positions as stated in the 1683 paper, assigning those in the meridians of Land's End (N) and the Pacific (S) to the internal sphere, and the other two to the crust (see fig. 3.3). This implied that neither dipole was antipodal; these ideas thus constitute the first isolated instance of a published phase-four hypothesis.[76]

The difference in style and content between the two postulates of 1683 and 1692 is notable. Part of the explanation may have been the preparation, printing, and publication of Newton's *Principia* during the period 1684–87, with which Halley had been closely involved. The shift in emphasis from a primarily descriptive approach toward a more analytical stance seems inspired by Newton's rigorous treatment of wide-ranging phenomena within the single framework of universal gravitation. Halley's proposed motions inside the Earth moreover conformed to Newtonian dynamics and celestial mechanics,

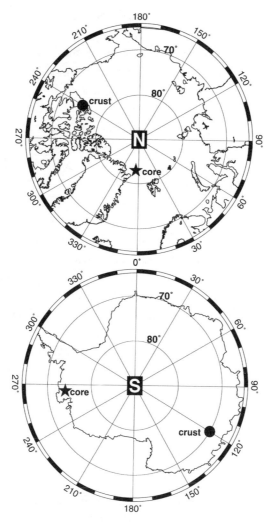

Fig. 3.3. Halley's nested double dipole theory (1692). Coordinates of the four magnetic poles in core and crust from his 1683 hypothesis were not assigned a depth until 1692. Double polar stereographic (equal-angle) projection.

in that none of the internal rotations would change the common center of the planet's mass. He was thus one of the first to apply to his own theories the physical constraints dictated by Newton's insights.[77]

Most of the response evoked by Halley's two geomagnetic hypotheses spanned the period from the 1720s to the 1760s. In 1721, William Whiston, proponent of a longitude scheme based on a single precessing dipole, quickly rejected Halley's fixed crustal magnetic poles. Twenty-eight years later, mathematician Leonhard Euler was able to prove mathematically that the observed patterns of declination could equally well be accounted for by a single disjointed dipole system, thereby dispensing with the need for a second dipole.[78] French critics primarily attacked Halley's isogonic charts. Among the more positive reactions was the extensive work done in the 1750s by Italian scholar Joanne Baptista Scarella, who tried to quantify Halley's system to determine contemporary positions of the four postulated poles. Furthermore, in the 1760s Dutch former master Meindert Semeyns extended Halley's concept to a fully quantified, nested triple dipole scheme.[79]

In retrospect, as far as development and spread of geomagnetic hypotheses are concerned, competing natural philosophies shaped the seventeenth century to a greater extent than in previous times. Gilbertian magnetic philosophy set the stage in 1600, a peculiar combination of animism, terrella mimesis, and painstaking laboratory research. The concept of magnetic rotation in celestial mechanics inspired Kepler to formulate his three laws of planetary motion, and gave the Jesuits an opportunity to attack heliocentrism. Cartesian doctrine meanwhile provided an alternative framework, in an attempt to reduce all of nature to principles of matter and motion. Corpuscular vortices proved particularly productive in the explanation of (geo)magnetism. Other forms of "particle physics" became the subject of research by Gassendians, English researchers from the 1640s, and the Jesuits from the 1670s.

Geomagnetic longitude schemes initially suffered a setback. Earlier simple notions of static dipoles and quadrupoles had fallen far short of expectation when confronted with the emerging, seemingly irregular patterns of observed declination. With the advent of overseas trading companies, missionary networks, and better instruments, it eventually became possible to compile hundreds of data points, where earlier only a few people had had access to more than a handful of scattered measurements. The single focus of magnetic attraction in inaccessible polar regions temporarily made way for global crustal heterogeneity; both continental landmasses and localized sources helped to explain a field exceeding dipole complexity. Cartesian vortices additionally cleared the way for atmospheric influences as a new disturbing factor.

The discovery of secular variation at first appeared to make matters worse, but in England it soon led to the transition to phase-three concepts: dynamic systems of tilted dipoles performing a circular precession around the Earth's spin axis in hundreds of years. Awareness of time-dependent change made all static notions redundant at a stroke, and provided an explanation for collections of undated observations previously deemed internally inconsistent. In a sense, theoreticians had to start over with a clean slate, much earlier work having become useless in the absence of recorded dates of the older measurements. At this time, around the 1640s, a temporary rift between English and Continental research appeared. Both managed to incorporate the idea of the geomagnetic field changing with time relatively easily, but along different lines. Various pairs of antonyms describe the contrast: experimental versus rationalist, atomistic versus Cartesian, quantitative versus qualitative, and reductionist versus complex. The Newtonian mathematization of the concept of force had little influence here; magnetism remained a part of the misty world of corpuscular effluvia.

In England, precessing dipoles, despite being brought into mathematical rule, were equally unable to adequately predict present or future field features. Logically, they eventually gave way to moving multipole systems, of which Halley's disjointed double dipole was the most sophisticated. It is important to stress that geomagnetism had become a subject to which investigators chose to devote substantial energy and time, not only to advance their own understanding, but also in direct competition. This trend is evident, for instance, in the early struggles for primacy between Bond, Phillippes, and Hooke, but also in Halley's questionable tactics in using Perkins's work, and the philosophical dispute between Hooke and Lister. The parallel research programs executed by the learned societies in London and Oxford also bear some characteristics of rivalry, growing out of the attempt to become the first to find proof of the corpuscular nature of magnetism. As far as geomagnetic hypotheses are concerned, the willingness to enter this particular arena may have been stimulated by the relative paucity of dated empirical evidence with which to support or reject new and existing claims. The inability of proponents to corroborate and of opponents to challenge postulates immediately upon presentation may very well have facilitated the development of new concepts, which would possibly have received short shrift if launched within the bounds of a rigidly defined theoretical discipline. It is therefore interesting to note how English seventeenth-century ideas still remained relatively similar, in all relying on one or two precessing dipoles.

Continental interpretations progressed at a different level, preferring a micro-mechanistic approach over a macro-mechanistic view. The composition of the deep Earth likewise evoked much speculation and debate: Galilei stressed the reigning intense pressures there, while Jesuits Kircher and Riccioli identified internal heat and chemical and physical action as potentially important local agents of change. Magnetic fibers and other quasi-organic notions, such as mutating bodies of iron ore, were moreover kept alive by several authors. Around the turn of the eighteenth century, Delisle undertook a more quantified but localized study of secular variation and acceleration. But despite this focus on change over time, all seventeenth-century Continental longitude schemes remained static, conceptual fossils incompatible with the emerged whirling world of magnetic circulation.

Once again, the question can be posed as to what extent the dynamics of the sketched development were driven by observations. For both Gilbert's and Descartes's interpretation of the Earth's field, deductive rationalism appears to have played a far more important role. The former's terrella proof of the planet's daily rotation was easily falsified by Grandamy, and even a small sample of declinations obtained near Asian, South African, or South American shores would have convincingly refuted the postulate of continental attraction of the compass needle. Furthermore, Cartesian philosophy relied heavily upon nature for providing phenomena to explain, but to a lesser extent for testing the validity of the proposed mechanisms. In Jesuit doctrine, some authors did invoke measured values, but these frequently remained undated, and primarily served to underline the absence of any pattern. Lastly, hypotheses pertaining to the deep Earth depended almost solely on analogy.

The discovery of secular variation provides an example of the opposite case. Series of measurements taken over a sufficiently long interval eventually showed a difference irreconcilable with existing notions of instrumental accuracy. The idea of the geomagnetic field's time-dependent change then spread quickly, and was reconfirmed time and again. In Borough's case, not all conditions defining a controlled measurement in 1622 had been met forty-two years earlier, leaving Gunter with a strong suspicion awaiting further corroboration. His untimely death offered Marr and Gellibrand the opportunity to finish the task a decade later. The continued London sequence also founded precessing dipole propositions by Bond, Phillippes, and Hooke. Bond's four examples of predicted inclination of course fell woefully short of a thorough test of the theory's robustness, which left Blackborrow with little difficulty in exposing its shortcomings. The Royal Society meanwhile undertook the first institutional compilation of magnetic data, instruction and encouragement of navigational practitioners, and archival research of ships' logbooks and travel accounts. Success was, however, very limited; Halley's 1683 quadrupole was still based on some fifty-odd points, few more than Plancius had used to derive his 1598 solution.

CHAPTER 4

The Age of Data

Geomagnetism in the Eighteenth Century

It was philosopher Immanuel Kant who coined the term *Enlightenment,* in an attempt to capture in a single word several eighteenth-century philosophical and scientific developments. To him it meant the reconciliation of observations (empiricism) and theory (rationality), culminating in a new understanding of the universe. Given certain fundamental concepts such as space, time, and causality, nature would show itself to be regular, harmonious, and subject to the power of human reason. Mathematicians such as Euler, Bernouilli, Mayer, Lagrange, and Laplace practiced this creed by developing new mathematics, and applying these successfully to celestial mechanics. Other examples include experimental physicists such as Gray, du Fay, and Franklin, who explored and described electrical phenomena; Linnaeus, de Buffon, and Lamarck, who classified the natural world; and Lavoisier, Cavendish, and Dalton, who replaced the earlier phlogiston theory of combustion with thermodynamics and atomic interaction. A deep sense of civilization's progress emerged, fostered in scientific circles and disseminated more widely through encyclopedias and lectures.

The eighteenth century also witnessed the rise of technology. The practical application of science led to many new machines, large numbers of craftsmen and engineers, the first polytechnic schools, the first iron bridges, and the modest beginnings of industrialization. It was the time when coal replaced wood in heating and iron smelting, and when gas lighting was introduced to some major cities. In the maritime realm, new instruments offered the navigator new opportunities and challenges: octant and sextant yielded more accurate determinations of latitude, the marine chronometer promised the same for the longitude, while taller visors significantly broadened the bearing compass's range in altitude and sighting opportunities. Commanders of exploratory voyages in the Pacific (such as Anson, Cook, Bougainville, Crozet, La Pérouse, and d'Entrecasteaux) gratefully implemented the available devices on board.

Meanwhile, ever more ships plied the Atlantic and Indian Oceans, among them those of the new French and Danish East India Companies, naval convoys, slavers, whalers, and the Hudson's Bay Company. This oceanic traffic not only transported unprecedented quantities of people and goods around the globe, but also generated a huge increase in the number of compass measurements made at sea. Processed by experts at home, the identified field features would lead to some radical revisions of earlier geomagnetic postulates. On the threshold of the eighteenth century, a closer look at data collection for scientific purposes therefore seems warranted.

Data Compilation

In France, the procurement of magnetic data for research purposes was long considered the domain and responsibility of scholars. Natural philosophers, mathematicians, and astronomers were all able to determine a meridian line and to perform careful measurements with sensitive equipment. At the French capital, conscientious efforts resulted in a substantial series of readings; the record of annual declination observations in and around Paris over the period 1657–1883 contains readings for all but twelve years. Around the mid–seventeenth century, some of this knowledge reached interested parties abroad through personal networks.[1] Nearer the turn of the eighteenth century, such information more often came out in print. Astronomical tracts, for instance, sometimes included statements regarding needle deflection at a site where heavenly phenomena were observed. Inasmuch as this scarce material allows generalization, the eighteenth century saw a broadening of the class of observers to include people more directly associated with navigation. Paris hydrographer Guillaume Delisle processed hundreds of logbooks of ocean voyages without questioning the professional competence of navigators. His port-based colleagues similarly came to be relied upon for supplying data obtained at regular intervals.[2]

Another increasingly important factor in the distribution of empirical data in the eighteenth century was the dissemination of the *Accounts of the Paris Academy of Sciences*. A sample of their contents over the period 1699–1760 reveals a large number of communicated observations of declinations, made on land, in ports, and during ocean voyages, by scholars, hydrographers, and navigators alike. In addition, the periodical paid attention to compass description, instrumental error, sighting techniques, possible technological improvements, inclination, secular variation, irregular field characteristics, and the geomagnetic determination of longitude.[3]

In the Dutch Republic, it was the Dutch East India Company (the VOC) that accumulated most geomagnetically relevant material. Navigators were obliged to hand in a copy of their logbooks after each voyage, to be examined for useful information by a company hydrographer or examiner of masters. Afterwards, the documents remained under lock and key, as their contents were deemed of strategic importance. Recorded compass behavior subsequently found its way into sailing directions, the accuracy of which required regular updates every few years. Several inventors of geomagnetic hypotheses in the Dutch Republic furthermore had a professional relationship with the VOC. The work of Plancius has already received attention. In the eighteenth century, longitude-finders Nicolaas Cruquius and Meindert Semeyns were company examiners, and thus had access to a large body of its primary sources. Regrettably, their published tracts feature very few of these observations. More extensive tables formed part of sundry navigation manuals, but most of the values listed there had been previously published elsewhere. Dutch natural philosophers such as van Musschenbroeck had to rely on the publications of the Royal Society at London and the Parisian Academy for the majority of their global geomagnetic data.[4]

The English case was somewhat different; there, land-based scholars fairly consistently encouraged navigators to supply them with data from faraway places. As early as 1582, master Thomas Bavin, about to set out on a quest to find the Northwest Passage to China, received an observation compass and an inclinometer, together with extensive instructions for marking the findings on a chart. During the first decades of the East India Company, William Boswell drafted a specific instruction for those masters who traveled beyond the equator. It advised them to determine a meridian on land, to which both compass and dip meter were to be applied, taking special care to avoid the effects of possible local deviation. In addition to declination, investigators in the early seventeenth century also paid specific attention to the relationship between inclination and latitude, and to magnetic intensity under the equator.[5]

The most concerted effort, launched from the 1660s onward, originated at the Royal Society. Requested information from various English and foreign parts, submitted by fellow researchers, regularly featured in the institution's correspondence.[6] Regarding the Society's own series of annual measurements at London, the earlier mentioned Magnetics Committee replaced its first-chosen location of the Whitehall dial with Gresham College in the mid-1680s. The site was relocated again in 1716, this time to the Society's yard at Crane Court, where Edmond Halley drew a new meridian in stone. Furthermore, apart from successive Royal Astronomers at Greenwich, other investigators also produced substantial series of measurements at London.[7]

Unfortunately, establishing durable bilateral communications with navigational practitioners met with less success. Neither the "Directions for Seamen, Bound for Far Voyages" (1665) nor the "Directions for Observations and Experiments to Be Made by Masters of Ships" (1667), in the *Philosophical Transactions*, provoked any substantial response. Peter Perkins's 1680 logbook survey would have been a valuable substitute, but it came to an abrupt end upon the compiler's death. Samuel Pepys proposed a similar initiative five years later, but to publication it never led.[8]

The natural philosophers' alternative to soliciting oceanic data from seafarers was to actively engage in magnetic surveys at sea themselves. During the period from 1600 to 1800, only two such scientific endeavors took place purely to advance geomagnetic studies, both by Halley around the turn of the eighteenth century. The story actually dates back to 1693, when Royal Society Fellow Benjamin Middleton sent the Society a proposal for an expedition which was to circumnavigate the globe to establish the proper coordinates of many landmarks and ports, and to take magnetic readings where possible. If a vessel of circa sixty tons could be procured and fitted out by the government, Middleton offered to cover all expenses for manning and victuals. Accordingly, the navy decided to build an eighty-ton pink, the *Paramore*, for that purpose. Apparently, even at this early stage, Halley was involved as well.[9]

By 1696 the ship was ready, but no more was heard of Middleton, and Halley was temporarily engaged in other business; so it was not until the autumn of 1698 that preparations for departure began in earnest. The objective had by then been scaled down to a ten-to-twelve-month naval survey of the North and South Atlantic. The astronomer received an appointment as captain, while his second in command was lieutenant Edward Harrison, himself a proponent of a geomagnetic longitude scheme. His hypothesis from 1696 was very similar to Halley's 1692 concept, and had not earned him favorable comments at the Royal Society. Harrison also questioned Halley's competence as head officer of a navy ship, even though the natural philosopher had had some previous sailing experience when he had traveled to St. Helena in 1676–78, to map the stars in the southern skies.[10] The frosty relationship did not augur well for the expedition's success, and so it turned out.

During the eight months after anchors were weighed in October 1698, the crew witnessed several clashes of authority between the two in their floating arena. It is unclear to what extent navigation suffered as a result, but the ship did end up in the doldrums, where so much time was lost that bad winter weather prevented exploration of the South Atlantic after arrival in Brazil. Thereupon, the voyage home was undertaken via the West Indies. After returning to London in June 1699, the ship underwent a refit, and Harrison was court-martialed for insubordination. But as the defendant had not directly and openly disobeyed any of Halley's orders, the only sustained charge was that of insolence, and Harrison got off with a reprimand. Afterwards, the lieutenant decided to rejoin the merchant marine.[11]

Halley had, in the meantime, marked his observations of the first voyage on a chart, which he showed to an assembly of the Society on 16 August 1699. Exactly one month later, the *Paramore* left port for a second attempt to survey the South Atlantic. This time around, its progress was only halted by icebergs around 52° south latitude, after which the expedition's track completed a large figure eight past St. Helena, Trinidad, Newfoundland, and back to Plymouth. Halley obtained about twice as many data points, which did, however, partially overlap with those of the first venture.[12] This dataset formed the basis for the world's first printed isogonic charts.

As regards the taking of inclination readings, the astronomer was regrettably far less diligent on his voyages. He had earlier carried a dip meter on his star-cataloging trip to St. Helena in the late 1670s, but had only taken two readings, both of them on land. He moreover failed to bring such an instrument on his magnetic surveys, probably because observation aboard ship remained troublesome and inaccurate. His compatriot astronomer James Pound, sailing to China with James Cunningham in 1699–1700, however, did take such readings. His data series spread via subsequent publication in the *Philosophical Transactions*. The French Jesuit missionary Feuillée followed soon after, traveling first to the East Indies in 1706, and around South America two years later. On both journeys he collected a sizeable number of high-quality inclination data.

A few similar compilations appeared in the second half of the eighteenth century. In the 1760s and 1770s the Swedish Royal Academy of Sciences provided Karl Gustav Ekeberg, a captain in the Swedish East India Company, with

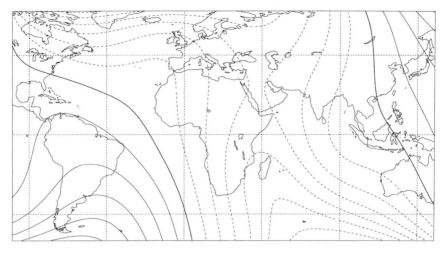

Fig. 4.1. Modeled magnetic declination at the Earth's surface in 1700. Bold curves = agonic lines; solid curves = northeasting; dotted curves = northwesting; contour interval = five degrees. Cylindrical equidistant projection.

inclinometers on several crossings to the East. London's Royal Society made similar arrangements for James Cook's circumnavigations, Constantine Phipps's Arctic exploration (1773), and Thomas Hutchins's 1774 expedition to Hudson's Bay. The last observer reported afterwards how he had twice set up the instrument on floating sheets of ice, to which his ship was temporarily grappled. The technique was not very successful, as the ice floes continued to drift and turn in the breeze, making alignment in the magnetic meridian difficult. Other sightings therefore took place only on shore.[13] In retrospect, the scanty evidence shows that the measurement of inclination at sea remained almost exclusively the prerogative of natural philosophers, despite the advent of better, standardized instruments.

At home, the study of inclination equally failed to rise to any prominence until the last quarter of the eighteenth century. The Paris sequence of observations is telling in this respect, featuring no more than eight values of I prior to 1780, of which only three are from the seventeenth century. The record for the city of London is somewhat better, with six reliable observations before the eighteenth century, and another thirty-two up to 1800. A rare instance of dip data cited in a nautical context can be found in Richard Waddington's *Epitome of Theoretical and Practical Navigation* (1777), which not only listed current figures for a number of English places, but also a series (1749–57) of both magnetic angles as observed by astronomer James Bradley at the Greenwich Royal Observatory.[14]

Quite a few researchers in the eighteenth century compiled long series of magnetic readings on land, and these led to a number of important findings, not least regarding the Earth's external magnetic field. Although the Earth itself is responsible for most of the surface magnetic field, it is by no means the only source. Through a variable flow of ionized particles, the Sun's magnetosphere interacts with the geomagnetic field to form several current systems around the planet, and these do affect surface conditions as well. In combination with the globe's rotation, magnetic measurements tend to vary slightly, both in the course of a day (*diurnal variation*) and a year. Nowadays, magnetic observatories keep constant track of these fluctuations, but in the eighteenth century it took sustained effort and dedication, firstly to establish their existence, and secondly to analyze their behavior. A few highly concentrated datasets have resulted.

The discovery of diurnal variation was made by London instrument maker George Graham. Although his speciality was clocks, he did not shrink from constructing sophisticated compasses accurate to two minutes of arc. In 1722, this able artisan had noticed a slight variability in their performance, initially attributed to pivotal friction. To investigate matters further, he deployed three twelve-inch needles, equipped with a lens to observe the smallest subdivisions on their finely graduated arcs. Graham first set up this *declinatorium* on the roof of his house in Fleet Street, where he had marked a geographical meridian, but wind disturbance soon forced him to move his experiments indoors. There he made over a thousand observations between February and May 1722, and discovered "that all the needles . . . would not only vary in their direction on different days but at different times on the same day, and this difference would sometimes amount upwards of 30′ in one day, sometimes in a few hours."[15]

Five years after Graham published his findings in the *Philosophical Transactions*, Dutch researcher van Musschenbroeck started to investigate the hypothesis that the needle's seasonal inconstancies were partly due to weather influences. Over the period 1729–31, he daily observed D and I at noon, and although he failed to find a significant correlation with atmospheric factors, he did determine that one of his needles' magnetization lasted about two and a half years before it became unreliable. Inspired by Graham's results, professor Anders Celsius and his colleague Olof Hiorter in Sweden decided to put diurnal variation to the test more thoroughly. In 1740–47 the two compiled over twenty thousand declination readings at Uppsala, corroborating the existence of the effect. A decade later, John Canton in England provided further proof with a series of over four thousand observations spanning the years 1756–57. He rightly suspected the cause to be solar, but resorted to the Earth's heating and cooling while it turned upon its axis to provide a plausible mechanism.[16]

In France, the Count de la Cépède contributed with six months of thrice-daily measurements in 1778–79, to confirm his postulate that the oscillations were the work of "the electrical fluidum." In an address to the Paris Academy and the navy, the author discussed a number of experiments, and claimed that the needle's susceptibility to nearby lightning could be counteracted by shielding it between two thick plates of glass. A decade later (1783–89), astronomer Jean-Jacques Cassini used freely suspended needles of about twelve inch in length. After some five and a half years of daily measurement he deduced a seasonal pattern changing every three to four months. Around the same time, Dutch professor of mathematics Jan Hendrik van Swinden embarked upon an even more strenuous research regime, taking readings every hour of every day for ten years.[17]

Whereas diurnal variation causes needle displacement of a few tens of minutes, the disturbance accompanying a magnetic storm can easily amount to a couple of degrees. In 1716 it was once again Halley who, in an account of a display of the northern lights over London, was the first to tentatively link the Earth's field with the aurora borealis. The luminous arch was noted to be always highest in the magnetic meridian, and the colored features in the sky seemed aligned with local inclination. Invoking a Cartesian double circulation of Gassendian atomic effluvia, Halley postulated "that this subtle matter . . . may now and then by the concourse of several causes very rarely coincident . . . be capable of producing a small degree of light."[18] He then went on to recall his nested-spheres hypothesis of 1692, and posited that the medium in between might be luminous, providing a perpetual day for life-forms that potentially inhabited the kernel. On very rare occasions, a small quantity of this substance could perhaps penetrate the outer shell and escape into the atmosphere. Ten years later he reiterated his belief that magnetic effluvia produced the display.[19]

It would take until the fifth of April 1741 before Graham in London and Celsius in Uppsala could confirm a direct geomagnetic relationship, after they simultaneously observed a magnetic storm that coincided with an aurora. The correlation was also observed several times in 1749–50 in Stockholm. Professor

Pehr Wargentin's findings there eventually became incorporated in a 1759 paper in the *Philosophical Transactions* by Canton, in which the author presented his own diurnal dataset. The English researcher had classified 29 of his 603 days of observation as irregular, but attributed these to the sudden generation of subterranean heat, affecting the electricity in the atmosphere above it. The preference of aurorae for northern regions he explained by proposing that the postulated increase in air temperature would be greatest there.[20]

An Eighteenth-Century Cross Section

Regarding the development and spread of geomagnetic postulates, Cartesian tenets continued to hold sway on the European mainland for much of the eighteenth century. The vortex theory of magnetic matter remained largely unchallenged, albeit somewhat modified. The most important shift in opinion concerned Descartes's original double circulation being replaced by a single stream. The mechanism was thought equally applicable to small magnetized needles and the entire planet. As hydrographer du Chatelard envisaged in 1735, the smaller system in and around a compass would become magnetically aligned in the larger global stream of corpuscles traveling between the Earth's poles. While Leonhard Euler considered their actual direction immaterial, Strasbourg physician and botanist Gilles-Augustin Bazin maintained the firm conviction that the magnetic effluvium exited at the north pole, and reentered at the south. He furthermore rejected the notion of the magnetic poles being able to move, attributing secular variation instead to slow "magnetic tides."[21]

A quite different circulation pattern emerged from the brain of Jesuit philosopher Sarrabat, who taught at the Grand Collège of Lyon. In a prize-winning essay (1727), he presented his geomagnetic vision, which was based on a data compilation spanning the previous twenty-six years, obtained from the Paris Academy and Jesuit sources. Sarrabat presumed that a spherical fire at the heart of the Earth was the driving force behind the expulsion of magnetic matter; a never-ending bombardment of particles assailed and partly penetrated the crust from the inside. Maximum flow surfaced at the magnetic equator, from which point it flowed northward or southward in the atmosphere, eventually going underground again at the poles. The internal fire supposedly rotated at a slight angle to the Earth's spin axis, creating a precessing dipole with a great circle agonic through the Azores and Canton, subject to westward drift. In explaining observed irregularity in declination patterns, the Jesuit philosopher thought that iron mines were of little influence. More substantial disturbances would arise from the atmospheric movements of air masses. He furthermore posited that crustal variability would affect the magnetic flow, which allowed stronger "exhalations" in some places than in others. This would give rise to substantial magnetic surface "currents," which were responsible for the deformation of the two agonics from their ideal meridional shape. One such current (in northwesterly direction) lay west of the prime meridian under the Tropic of Cancer, another (flowing southwest) assumedly roamed southeast of China. According to Sarrabat, future empirical research should concentrate on westward drift, the position of the agonics, and location and force of the magnetic currents. The ultimate aim was to gain sufficient insight into the dynamics at work to be able to draw up annual global tables of predicted declination.[22]

Only a minority of scholars actively opposed the Cartesian rationale for (geo)magnetism. Astronomer Pierre-Charles Le Monnier in 1733 still adhered to the concept of moving subtle magnetic matter, but at the same time questioned the principle of vortex circulation. More outspoken was van Musschenbroeck in his 1739 *Essays on Physics*. But although he explicitly rejected Cartesian doctrine, he failed to produce a viable alternative. In 1760, astronomer and mathematician Tobias Mayer considered Cartesianism useless and inept; physicists Aepinus, Lambert, and Coulomb later joined him in this opinion. Meanwhile, Kircher's idea of magnetic fibers gained new support in the 1740s from Daniel Bernoulli, who reiterated it in altered form. The mathematician assumed all magnets to be permeated by channels much like blood vessels. These transported the magnetic effluvium in two opposing directions by constant undulations through a system of valves. Bernoulli's terrestrial interpretation remained free of quantified constraints. His method to find longitude therefore depended not on predictions of polar position, but on an envisaged empirically derived chart, combining isogonics and isoclinics. Then "the intersections of these two kinds of curves could perhaps shed some light on one's location."[23]

For England, Newtonianism has frequently been the umbrella term with which to classify and explain all natural phenomena, evoking connotations of the *Principia*'s rigorous approach to dynamics and celestial mechanics. But although this mode of thought came to replace Continental Cartesian tenets in these two subdisciplines from the 1740s, it would be a misleading generalization to give all of experimental physics (such as practiced in the Dutch Republic by Boerhaave, 's Gravenzande, and van Musschenbroeck, and by Brisson, du

Fay, and Nollet in France) such a Newtonian label. Inasmuch as Newton's work was an inspiration there, it was not through the *Principia*'s forces acting at a distance, but rather the effluvia pervading his other great work, the *Opticks*. Nowhere is this more apparent than in the study of magnetism. Both Newton himself and many of his disciples continued to adhere to magnetic attraction by mechanical impulse of circulating material effluvia.[24] During the second half of the eighteenth century, these ideas became subsumed in more general theories of ethers, already relied upon to transfer potentially related phenomena such as electricity and light. By this time, the distant action of gravity had been uniformly accepted to rule the solar system. This was in sharp contrast with the ethereal medium, still assumed a necessary contact scaffolding for electromagnetic effects throughout the nineteenth century. Even after repeated failures to detect the ether in the 1880s, and Albert Einstein's classic paper on special relativity in 1905, which altogether dispensed with the need for such a carrier, some physicists continued to search for it until the 1920s.

Geomagnetic hypotheses in the eighteenth century ran the gamut of existing explanatory schemes, some hardly distinct from predecessors over a hundred years older, others adding surprising new features. Each phase counted proponents in England, France, and the Dutch Republic, occasionally reinforced from other countries. Yet communication seems to have been primarily on the level of exchange of datasets; only rarely does one find direct mention of contemporary (or slightly earlier) competitors. Researchers still seem to have worked in relative isolation. Perhaps the sizeable reward issued by the English Board of Longitude in 1714 was influential in this respect; each chicken wants to lay the golden egg herself.

Unfortunately for their promulgators, many of these eggs were long past their sell-by date. Even schemes concerning fixed antipodal pole pairs staged comebacks time and again. It is curious to find such remarkable longevity in an idea so fundamentally flawed as the static tilted dipole, despite repeated confirmation of Gellibrand's findings of change over time at London, and ubiquitous references to secular variation in academic and maritime publications. In defense of these hopefuls, one could argue that many were amateurs, being neither seasoned navigators nor hydrographers nor natural philosophers well versed in the relevant sources.

This makes the time-invariant hypothesis put forward in 1738 by Nicolaas Cruquius all the more curious. The Dutch surveyor, cartographer, engineer, and Fellow of the Royal Society from 1724 fulfilled the office of examiner of masters at the Delft chamber of the VOC between 1725 and 1739. In this capacity he had access to all navigational logbooks written by officers sailing with East Indiamen from that city. Furthermore, his private notebooks, in which he collected the oldest Dutch series of meteorological observations (1706–1734), contain scattered sequences of secular variation, recorded at various locations the world over. It remains a mystery why, in spite of this evidence, he chose to publish a longitude scheme in 1738 which in complexity is more akin to sixteenth-century Portuguese solutions than most contemporary alternatives. Although he shielded the mathematical foundation from public scrutiny, his four tables of gridded declination predictions barely disguise complete symmetry relative to a single agonic great circle. In the accompanying instructions, Cruquius advised the mariner of a twofold application (assuming the latitude to be known): given longitude to find declination, and given declination to find longitude. Fortunately for seafarers, no trace of practical implementation has come to light.[25]

Eighteenth-century antipodal dynamic dipoles also continued to make new appearances, mainly in the 1730s. Two examples will have to suffice here, designed respectively by Oxford physicist Servington Savery (FRS) and Jean-Philippe de la Croix, a shipboard scribe in the French Navy. The latter's proposal, touted for over a decade and a half, was fairly preliminary in nature; at no stage in its development did it aspire to quantify particular parameters. Its history started in 1731, when the inventor compiled a list of 516 declinations. In spite of the lack of any obvious pattern, the French officer submitted two longitude schemes to the Paris Academy that same year, one supposing spatial regularity in D, the other in I. At the heart of both lay an antipodal tilted dipole, with unknown colatitude. De la Croix may have attempted to locate the magnetic pole by means of a cross-bearing, but the outcome has remained obscure. Académie examiners Cassini and Maraldi were quick to reject the declination-longitude method on the basis of the erratic isogonics on Halley's charts. They went on to raise objections against the second scheme too, noting that it would necessitate inclinometer readings at sea, and that practical benefit would only ensue if the dipole's tilt was found to be substantial. Nevertheless, they subsequently endorsed further experiments.[26]

In 1737 de la Croix wrote to Secretary of State de Maurepas, suggesting that dip observations be made at Brest and Strasbourg. Over the next two years, naval administrator de Mayran coordinated communication concerning vertical needle measurements, which were to be collected at Marseille, Nantes,

Brest, and Bayonne. A plan for simultaneous observation of inclination near Perpignan and Toulouse (assumed to be on the same meridian) was made in 1740, with unspecified results. The following year, de la Croix had his dip-longitude method reevaluated by a committee of three at the Académie: Cassini, de Fouchy, and Saivegure. They supported the idea of a limited survey of dip, judging it "not only useful for the verification of the method of M. Delacroix, but also for physics and navigation."[27]

Emboldened by this official backing, de la Croix developed a more ambitious project in 1743, entailing a direct astronomical fix of the magnetic equator at two distant points, followed by establishing the position of each degree of magnetic latitude. The Académie's reaction in 1744 was hesitant: since, in all likelihood, more than two magnetic poles existed, the survey would have to be much wider. De la Croix accordingly changed tack again two years later, when he informed de Maurepas that the only French territories crossed by the magnetic equator were the French East Indies. Apparently, the protagonist had already arranged cooperation with the Compagnie des Indes. He launched an alternative plan of investigation in 1747, involving inclinations gathered on Mediterranean and Atlantic coasts from Constantinople to North America. Notwithstanding this sustained lobby by its eager proponent, it appears that no such attempt was actually undertaken; the project unceremoniously died, and was buried among hundreds of other naval files.[28]

Across the Channel, Servington Savery added insights to the body of geomagnetic thought in 1732 (see fig. 4.2). These had little to do with his own, rather mundane precessing dipole concept, which closely resembled Bond's. More interestingly, he was fully aware of Halley's legacy, but rejected the double dipole shell theory on the grounds that the nearness of two poles of like denomination would repulse one another to a considerable degree. This would cause a latitudinal shift in nucleus and shell, an effect that observations did not support. In addition, he assumed tidal forces from Moon and Sun to affect the freely suspended inner core. He tested this hypothesis by means of Graham's sequence of diurnal variation. Unfortunately for him, these fluctuations followed the solar, rather than the lunar day. He nevertheless held on to the conviction that "there is a central magnet loose from the Earth revolving on an axis parallel to that of the Earth in nearly the same time the Earth does."[29] Remarkably, his new explanation for observed declinations at odds with the predicted pattern was "from irregularity's [sic] on the surface of the internal Magnet, whose mountains may attract somew.t stronger than its other parts."[30]

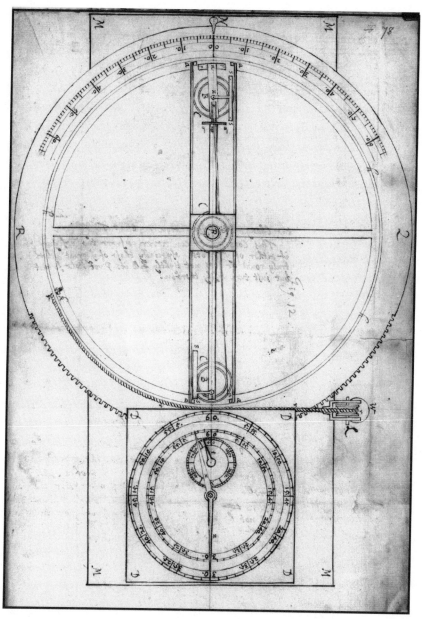

Fig. 4.2. Savery's sketch (BL Add. MSS 4433 fol. 78) of his proposed invention (1732) of a magnetic instrument capable of measuring core topography. By permission of the British Library, London.

Elsewhere, he described these features as "protuberances," quite probably the earliest recorded supposition (1732) of core topography affecting the surface field. He even intended to develop an instrument to locally measure the depth of the internal magnet's surface.[31]

Before discussing disjointed dipole solutions (the fourth phase of geomagnetic hypotheses), two exceptional precessing dipoles and a simple quadrupole deserve mentioning. The first of these was conceived by Welsh physician Zachariah Williams, who spent over twenty-five years trying in vain to sell his idea, without ever disclosing the defining characteristics. After moving from Wales to London, he made an excellent start; he met Royal Society Fellows Halley, Whiston, Hauksbee, Desaguliers, and Molyneux, presented magnetic experiments and excerpts from his prediction tables at the Royal Society, and obtained the lasting patronage of Samuel Johnson.[32] His quest for "suitable and seasonable encouragements," however, suffered severely from accusing Molyneux of intended plagiarism, and refusing to have the Admiralty test his method at sea, as Molyneux had suggested prior to his death. Williams had initially coaxed various investors into keeping him financially afloat with promised shares in the imminently expected reward. But as the years passed, restlessness grew among them, and eventually they wrote him off. The longitude-finder ended up in the charterhouse for the poor, but still managed to allocate funds for modest publications full of spite, melancholy, and updated magnetic forecasts. The underlying mathematical rationale he took with him in his grave. Nevertheless, the values generated for eleven European cities over the interval 1660–1860 proved sufficient to enable a full reconstruction (see fig. 4.3). The system turns out to be unique in having the dipole perform an *elliptical* orbit in about 591 years. This well-kept secret obviously had no influence on contemporary geomagnetic interpretations by others.[33]

The second exceptional dynamic dipole concept came from Germany. It was the intellectual property of Tobias Mayer, well known for his lunar tables, which helped to solve the longitude problem. In an unpublished tract of 1760 (*Theoria Magnetis*), he proposed a magnetic dipole of infinitesimal size, placed about one seventh of the Earth's radius distant from the planet's center, at 17° N and 159° W of Ferro (Canaries). In addition to precession and displacement toward the crust, its tilt was also subject to change; the dipole's designated co-latitude (then 11°30′) increased annually by eight minutes and fifteen seconds of arc, or so its author believed.[34]

Lastly, Arnold Maasdorp, a Dutch "mathematical amateur," as he intro-

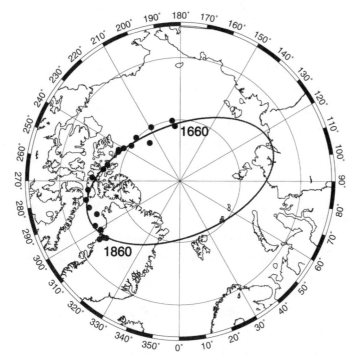

Fig. 4.3. The Williams hypothesis. Estimated dipole positions are plotted for each decade between 1660 and 1860; the best-fitting elliptical path is superimposed. Polar stereographic (equal-angle) projection.

duced himself, living in Cape Town (South Africa), designed a dynamic quadrupole system in 1753. In that year he published a booklet in Amsterdam in which he held slight irregularities in the Earth's daily rotation responsible for declination and secular variation. The author considered the field itself perfectly symmetrical; two perpendicular agonic great circles divided the globe in four equal parts of alternating sign, with maxima midway between each pair of adjacent agonic meridians. Sadly, not a single declination value was supplied; these would still have to be gathered in a detailed survey of one of the quarters. Nor did the author place the Atlantic agonic prime meridian at a specific longitude. He imagined the whole system to rotate relative to the crust in regular fashion, but even the direction had yet to be established. Nevertheless, Maasdorp assured his audience that the longitude problem was practically solved.[35]

Disjointed Dipoles

The acceptance of declination as a real phenomenon had instigated the transition from axial to tilted dipoles. The realization that agonics did not form great circles had brought about the adoption of multipole constructs. The discovery of secular variation had introduced the concept of time-dependent field change. But still reality failed to fit the corset of theory to an adequate degree. Philosophers had advanced a multitude of near-surface disturbances to explain residual features, but their very nature made them unsuitable for quantification and prediction, leaving lingering doubts as to whether any of the current interpretations was correct. Each step forward had increased complexity by requiring that more parameters be evaluated. Could other factors still be awaiting discovery? If orientation, number, and movement of pole pairs were insufficient, which remaining aspect needed its simplifying yoke lifting?

The fourth phase of complexity in geomagnetic field representation constituted the abandonment of magnetic poles in diametrical opposition. It was different from the previous three in gradually gaining ground over decades, from uncommented coordinates via mathematical proof to explicit and extensive treatment of each pole's habits. The bond between magnetic north and south was not severed with a single blow, but slowly stretched like an elastic band. At the same time, notions of solid cores spinning inside enveloping shells came under pressure. Propositions that the internal magnet was merely somewhat irregular in shape could still explain a fixed difference in longitude other than 180 degrees between the two concentrations of magnetic force. But by the 1790s several researchers had separated polar *motion* as well. And just as the magnetic mountain had earlier been found incapable of geographical displacement at the surface, a solid magnetic kernel seemed unable to accommodate "observed" independent polar behavior. This seems to be the main reason why phase-four hypotheses tended to concentrate on field description, rather than supplying a causal mechanism. In the following final two sections, the most notable of these systems will pass review.

Few individuals can justifiably claim to have made a more distinct mark on the early history of geomagnetism than Edmond Halley. His visionary spinning kernel provided inspiration for many, he was a driving force behind the first magnetic survey at sea, he introduced printed isogonic charts, and he wrote on the geomagnetic origin of the northern lights. In addition, an often-overlooked trait of his 1692 conjecture was the fact that neither of his dipoles was antipodal. The longitudinal distance between the poles near Land's End and Celebes was about 125 degrees, and their colatitudes differed by 13 degrees; the crustal Pacific and Californian poles were less than 30 degrees apart when measured along a parallel. Halley did not stress this fact himself, concentrating as he did on terrestrial mechanics.

A second instance of near-mute admission of a disjointed dipole was Whiston's postulate for the south magnetic pole (1721). According to the sober description, it was ring-shaped and resided in the Indian Ocean, some ninety degrees east of its northern neighbor, and at more than double the latter's colatitude. Nowhere did the author express awareness of its novelty, its lack of natural analogues, or the need for rigorous mathematical description.[36] Yet another disjointed polar arrangement came forth in 1734, generated by Swedish natural philosopher and mystic Emanuel Swedenborg. His two poles pursued circular paths eastward at a colatitude of 22°30′, induced by Cartesian magnetic vortices that pervaded the solar system, oriented in the plane of the ecliptic. In 1720 both poles temporarily resided in the western hemisphere (in Hudson's Bay and the South Pacific, at a longitudinal distance of 33°30′). The northern one required 386 years to complete one orbit, whereas its southern counterpart took almost three times as long (1,080 years). Despite a large number of comparisons made between observed and expected declination at various sites, Swedenborg, like his predecessors, failed to elaborate on his rejection of dipoles in diametrical opposition, either from first principles or deduced from empirical results.[37]

It was precisely in this last respect that Leonhard Euler finally supplied a more thorough analysis in 1757. In his *Recherches sur la Déclinaison de l'Aiguille Aimantée* he set out to prove mathematically that declination patterns as exhibited by contemporary isogonic charts did not necessarily have to arise from the interaction of four magnetic poles. As it turned out, the only type of dipole capable of producing a similar distribution had disjointed poles. This finding was in accordance with the scholar's rejection of a separate magnetic entity in the deep interior; instead he deemed mutating crustal iron and lodestone deposits of sufficient force to account for all of the Earth's field.[38]

While Euler thus largely rejected Halley's construct, a Dutch East India Company examiner of masters had meanwhile taken the opposite step of expanding it further. Meindert Semeyns of Enkhuizen claimed to have developed his triple dipole (in nucleus, intermediary shell, and crust) over a period of forty-five years, and submitted it to various official bodies for approval. Extensive correspondence and several publications from the 1760s make it one of the best-documented Dutch schemes available. Although he did not think highly of Euler's crustal dipole, the former navigator did tacitly undersign the notion of disjointed dipoles; the longitudinal distance between each of his pairs of poles amounted (from the inside out) to 90, 177, and 104 degrees respectively.[39]

Barely two decades later (1788), French academic Georges-Louis Le Clerc, better known as the Count of Buffon, inferred a crustal quadrupole from the observed presence of three agonic curves. He considered the actual location of all poles to be subject to unpredictable change, due to earthquakes, eruptions of molten rock, and the mining of iron ore.[40] Instead of giving their imagined coordinates, he merely tabulated observations of D and I, ordered for practical navigational application by value and by latitude.

A more theoretical approach was evident in the work of compass maker Ralph Walker of Jamaica. His 1794 *Treatise of Magnetism* was a mixture of tabulated measurements (for practical purposes), advertisement for his novel compass, and geomagnetic musings. He perceived two poles, moving from east to west without any regularity. This he deduced from his dataset, the general westerly displacement of the two identified agonics, and the principle that "the Earth or any other moving body cannot give a motion greater nor even equal to what it is itself possessed of."[41] His tenets represent an uneasy marriage of Gilbertian and Cartesian ideas. He envisaged corpuscular flow to emanate from each pole towards the other through the atmosphere. The planet having an oblate shape, the discharge of magnetic matter at one end and its reception at the other would take place some distance from geographic north. These points would have been in opposite meridians, were it not for mutual polar attraction drawing them closer together. This pull was itself counteracted by the magnetic fluid adhering to the Earth's solid surface; the actual location then resulted from the changing balance between these forces. Gilbertian continents, believed to be more magnetic than the oceans, not only locally diverted the flow, but also acted as a global drag factor on both poles as they passed by. The combined influence of these constantly changing factors precluded the establishment of any fixed period of revolution for either pole. At the time of writing, Walker assessed the colatitudinal difference to amount to six degrees, and 150 degrees in longitude.[42]

The last proponent of geomagnetic hypotheses to receive attention here is John Churchman. Over a period of seventeen years, this American surveyor released no fewer than three disjointed dipole concepts into the public domain, in addition to an isogonic globe, and four editions of his *Magnetic Atlas*. These endeavors deserve a few paragraphs for a number of reasons. For one, the archives of the English Board of Longitude hold sizeable files on his efforts, which permit a detailed reconstruction of events. For another, hardly any historian of science has so far taken the trouble to discuss his correspondence and publications at length. Thirdly, the presented postulates themselves display a modest evolution, which exemplifies and magnifies the issues involved around the turn of the nineteenth century. Last but not least, Churchman's tale is a good story.

Hope and Glory

Southeast of Philadelphia, the state border of Pennsylvania runs straight along a parallel of latitude.[43] In 1702 John Churchman's great-grandfather settled on a tract of land there called Nottingham (in the county of Chester). It lay just north of the state line, and was parceled out in square lots of five hundred acres each, all neatly aligned with the artificial divide. However, a land survey in 1736 confirmed this line at a slightly different angle. As a result, it then cut Churchman's land in two; the northern part still lay in Pennsylvania, but the southern tip had become the latest addition to Maryland. The parallel had been laid out by course and distance, with bordering trees being notched, and their position and date of marking recorded. Each state's land office moreover kept a copy of the relevant data for future reference. These clearly showed that the trusted compass bearings relied upon in surveys were constantly shifting away from the delimiters carved out in the tree-lined landscape.

John Churchman was very much aware of this fact. His grandfather, his father, and he himself had all held commissions as surveyors under the proprietors of Pennsylvania. Armed with his experience and the notes of his predecessors, he considered himself in an excellent position to establish a local estimate of the nearest magnetic pole's wanderings. Having heard of the Board of Longitude's reward, he set out in the late 1780s upon a venture that would keep on rolling into the next century. His mathematical and cartographic skills would serve him well in this undertaking. Part of his success may have been due to his multipronged approach: not only did he draw isogonics for charts and globes, he also gathered and made field observations, repeatedly calculated the dipole's twisted position, attempted to compile the world's first *Magnetic*

Almanac, and bombarded official and scientific bodies with letters and memorials. He may have been primarily commercially motivated; he surely was a driven man.

Churchman's first memorial to the board (14 February 1787) heralded the author's discovery of fixed principles in magnetism. These would ascertain to great precision the longitude of places in all parts of the globe, based on the exact delineation of the magnetic poles' position. At the time, Churchman still thought polar motions equable and uniform from west to east. In a follow-up letter (sent 31 March), he went into more detail, rendering coordinates for the dipole as it presumedly had stood in 1777. The arrangement was clearly disjointed (exhibiting 4° difference in colatitude, and 135° distance in longitude); the north magnetic pole revolved in little under 464 years (see table 4.1). Churchman had sketched the perceived isogonic distribution for 1777 on a globe, and claimed that it agreed well with Captain Cook's magnetic observations. Together with a separate table predicting half a revolution in the northern hemisphere (for 1657–1888), the surveyor shipped the whole to England for expert approval. Via several intermediaries all items eventually reached Henry Parker, then secretary of the Board of Longitude. The following year, this official passed them on to the Astronomer Royal, Nevil Maskelyne, for further study. Meanwhile, the inventor had announced "a proposal of publishing by subscription a variation chart on a new plan," which attracted a lot of attention. Churchman also tried in vain to enlist the cooperation of London mathematical practitioner George Adams. The American wanted him to produce a higher-quality example of his roughly executed globe, with the intention of obtaining an exclusive privilege for vending these in quantity, marketing them as longitude-finding aids at sea.[44]

In 1789, while the former surveyor was busily at work on his chart, and calculating declinations for a planned geomagnetic almanac, the United States Congress became involved. Churchman had petitioned the House with a detailed description of his analysis and its benefits for determining longitude. In addition, he had proposed that the federal government fund an expedition to Baffin's Bay to find the north pole's exact coordinates. After a personal appearance of the petitioner before a special committee, Congress decided "that such efforts deserve encouragement, and that a law should pass to secure to Mr. Churchman for a term of years the exclusive pecuniary emolument to be derived from the publication of these several inventions." This also inspired a separate motion for a more general bill, "securing to authors and inventors the

Table 4.1 Churchman's Three Geomagnetic Hypotheses

Year	Pole (yr)	Colatitude	Longitude	Relative Orbit	Period
1787	N (1779)	13°56′	274°48′ E	counterclockwise	464 years
	S (1777)	18°	140° E	unknown	unknown
1790	N (1777)	13°56′	269°02′ E	counterclockwise	426 years
	S (1777)	18°	140° E	clockwise	5,459 years
1794	N (1794)	30°55′	225° E	counterclockwise	1,096 years
	S (1793)	25°14′	158°50′ E	clockwise	2,289 years

Source: CUL, RGO/14/42 no. 5 (1787) fol. 60–62: Churchman, *Magnetic Atlas* (1790), 94–95, 105–6; Churchman, *Magnetic Atlas* (1794), 34–45.

Note: N / S = North / South magnetic pole (year of stated position). All longitudes are relative to Greenwich.

exclusive right of their respective writings and discoveries." A geomagnetic hypothesis was thus instrumental in the founding of U.S. copyright and patent law. Financial concerns regrettably prohibited sending ships out on scientific missions at the time, or so Churchman was told.[45]

While these affairs were conducted, Parker had finally conveyed the board's judgment of the American's longitude scheme. The main criticism concerned the fact that Leonhard Euler had already put forward a disjointed dipole solution in 1757. Churchman countered in a printed address, to the board and several learned societies in Europe and America. Among the points he raised was the fact that Euler had not explicitly applied his system for the determination of longitude, and had similarly failed to calculate polar positions over time. The inventor had also heard rumors of clandestine copies of his globe. In his accompanying letter he claimed priority for the production of isogonic globes, and offered to come to England to support his propositions in person.[46]

The following year (1790), the first edition of the *Magnetic Atlas* appeared in print. It contained an extensive theoretical discourse, which founded Churchman's second geomagnetic hypothesis, and two new tables predicting the path of each pole, both annually (1459–1884) and per century (3923 B.C. to A.D. 3877). Their orbits he presumed to be circular, but an elliptical path was noted as a possible alternative. In comparison with the previous conjecture, specifications pertaining to the southern half of the world were now more fleshed out. The longitudinal distance between the two poles, as they had stood in 1777, amounted to 129 degrees; their colatitude remained unaltered. The author imagined the dipole's precession to be eastwards in an absolute sense, but relative to the Earth's surface, the south pole lagged behind, taking some 5,459 years to complete a clockwise revolution, against 426 for its (not quite) opposite number, turning counterclockwise (see table 4.1).

The *Magnetic Almanac* was once more promised to be soon completed. Since Churchman made no mention of it ever after, the work was presumably abandoned prematurely, perhaps due to revised parameters requiring the surveyor to start anew one time too many. The separate geomagnetic chart consisted of twelve sections which, when pasted onto a sphere, would cover the northern hemisphere with the isogonics for 1790 traced upon them. Formal presentation to the board and the Admiralty took place on the first of October that year. It was also a commercial success, listing 288 subscribers willing to advance part of the printing costs. In a separate, slightly earlier letter to the board, the longitude-finder furthermore took the opportunity to ask the board's members to convince the Royal Navy to fund his intended expedition to Baffin's Bay. The objective remained the same: to exactly pinpoint the location of the northern magnetic pole.[47]

The following year passed by uneventfully, but in 1792 a new proposal invited potential customers to subscribe to a second edition of the atlas. This time around, the leaflet promised, the charts would not only cover both hemispheres, but also include lines of equal inclination. In a memorial to the board, Churchman once more returned to his Arctic pet project with the political argument that an expedition to Baffin's Bay should sail from England, since the territory fell within its dominions. The trip was this time thought necessary to quantify the exact rate of polar movement. However, the navy failed to be smitten by the idea.[48]

From 1793 the ex-surveyor engaged in extensive traveling. He first took magnetic measurements at various sites along the North American east coast, and then decided that his cause would be best served by moving temporarily to London. Once there, he announced to the board the imminent arrival of the second edition of the atlas, suggesting that a number of copies be purchased to facilitate extensive study of the theoretical principles. As promised, the *Magnetic Atlas or Variation Charts of the Whole Terraqueous Globe* duly appeared in print, incorporating two hemispherical projections at the back of the book. Isogonics for the year 1794 traced declination at five-degree intervals, while isoclinic curves, advertised as a supplementary aid in finding longitude, delineated 0, 30, 60, and 80 degrees of dip. The description furthermore contained

Churchman's third geomagnetic hypothesis, likewise prepared for the year 1794, and partly built upon his own observations. This time, the tireless proponent had substantially changed the polar coordinates, and not just on account of secular variation (see table 4.1). Colatitudes had increased from about 14 and 18 to 31 and 25 degrees, while periods of polar revolution had changed to 1,096 years (north) and 2,289 years (south) respectively. A table of observed and predicted declination at London since 1622 moreover showed a good fit. A separate memorial to the board continued to promote the idea of Arctic explorations.[49]

Subsequent editions of the atlas appeared in 1800 and 1804, but little new was added; the main body of the text continued to maintain the 1794 interpretation almost word for word. The third and fourth version of the charts did feature the novelty of having dotted isogonics for each degree, thereby reducing the need for visual interpolation. In the memorial that offered the third edition to the board, Churchman shrewdly returned to the question of primacy with respect to Euler's efforts, this time in a more assertive tone: "If professor Euler by discovering the number of magnetic poles or points, discovered the magnetic method, might it not be supposed that he who first discovered the Sun and Moon discovered the lunar method? But the lunar method was of but little use before the Nautical Almanac was planned by that great astronomer doctor Maskelyne, and printed under the direction of the Board of Longitude, whose publications have no doubt been the happy means of giving to the British Navy that superiority for which it is universally famed." He also pleaded to the members to consider "whether or not so much labour may be worthy of any reward."[50]

One of the learned societies Churchman had sent his tracts to was the Russian Imperial Academy of Sciences. Before the third edition of the atlas came out, the author had become a fellow there. In 1803 the opportunity arose to visit St. Petersburg and present the latest work to the Russian Admiralty in person. It was from this port that Churchman wrote to the Board of Longitude on 24 April with news of his latest invention, a magnetic instrument capable of indicating declination and dip at the same instant. It consisted of a three-inch terrella floating in a bowl of mercury underneath a universal sundial. When properly oriented in the meridian, the device's magnetic direction in the horizontal and the vertical plane could be immediately read off. Naturally, it remained to be seen whether the device could be kept steady enough to function at sea. In the letter, Churchman also reduced expectations regarding the determination of longitude by means of inclination, which he had found upon closer inspection to be not altogether regularly distributed, and almost constant over some large parallel tracks of ocean. Instead, he suggested it might be usefully employed to correct latitude in dark and cloudy weather, when no other observations could be made. Perhaps he was unaware that this idea had already proved fatally flawed two centuries earlier.[51]

But worse was yet to come. In 1804, the very year that the fourth edition of the *Magnetic Atlas* came out, Churchman finally gained more insight into the field's true complexity. Having returned to London, he had consulted a chart of Baffin's Bay made by Aaron Arrowsmith, which carried recent magnetic measurements there. The ultimate conclusion drawn was a devastating one. After recalling Euler's work supporting the notion that two disjointed poles would suffice to explain the observed field, Churchman had to admit to the board: "I was inclined for some time to believe the same, but since more . . . mature deliberation and from a multitude of observations it appears that two alone are not sufficient."[52] Truth was only to be found in the empirical domain, and isogonic charts would always require new and extensive magnetic surveys, rather than carefully calculated polar paths stretching thousands of years into the future. The American therefore offered to participate in further experiments on a voyage around Britain. But the fire had gone, as had the belief that disjointed dipoles held the key to formalized field description. With all his hopes extinguished, perhaps it was a mercy that the Philadelphia surveyor expired at sea while homeward bound.[53]

If the sixteenth century was a formative period full of diversity, and the seventeenth century an era dominated by competing theoretical interpretations, the eighteenth century is perhaps best described as the age of magnetic data. Never before were such wide expanses of ocean traversed by so many, carrying such accurate instruments made by specialized craftsmen. Never before had the measurements obtained therewith been subjected to more scrutiny from individuals and institutions. Compilations easily ran into many thousands of data points, enabling better coverage and reliability of observed field features. Their publication made them available to the learned and the curious alike, with the longitude problem and its associated rewards as a hidden or overt motivation. Scholars invested time and effort to gather sustained series of readings at a single location, first to track secular variation, later also to investigate daily and yearly trends, as well as the correlation with the northern lights. The theoreti-

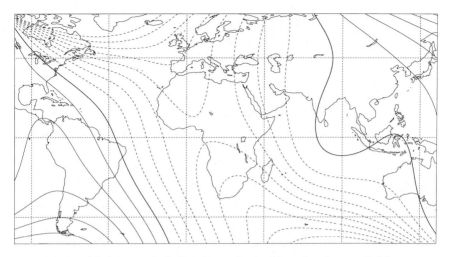

Fig. 4.4. Modeled magnetic declination at the Earth's surface in 1800. Bold curves = agonic lines; solid curves = northeasting; dotted curves = northwesting; contour interval = five degrees. Cylindrical equidistant projection.

cal foundations underlying new postulates reached a certain level of consensus on both sides of the Channel, despite superficial differences. Magnetic force was deemed effluvial in nature, and the former focus on magnetic particles came to be replaced by the notion of an all-permeating ether. Nevertheless, some proponents kept remnants of older ideas alive, and substantial controversy remained regarding the actual distribution and direction of magnetic flow. With a few notable exceptions, the concept of an internal lodestone kernel became more peripheral. This development displays a general shift in emphasis from causal theories toward descriptive hypotheses. Furthermore, national boundaries do not seem to have had a restricting influence on the spread of geomagnetic concepts at all: Halley's work was promulgated by Scarella in Italy and extended by Semeyns in the Dutch Republic; Swedenborg in Sweden drew freely on the intellectual capital of both Descartes and Halley; Euler's foundations supporting disjointed dipoles even reached Churchman in North America.

As far as chronology is concerned, a time lag exists between the adoption of postulates of any given phase of complexity by experts and by amateurs. Whereas static tilted dipoles had been almost exclusively the domain of seasoned Iberian navigators in the sixteenth century, these conjectures had two hundred years later been relegated to a second league of amateurs. The picture is more varied for dynamic dipoles and multipoles, presumably because phase-three concepts were going through a transition in this respect; individuals of diverse backgrounds still adhered to their underlying regularity. The relentless pressure of new datasets meanwhile forced scholars to seriously consider disjointed dipoles. Initially, schemes merely tacitly acknowledged that observations had become physically incompatible with a pair of poles in diametrical opposition. Only around the second half of the eighteenth century did Euler give phase four a more mathematical footing.

A second temporal aspect deserving attention is the shortening of each consecutive phase's period of acceptance. Axial dipoles had been a trusted frame of reference since Peregrinus in the thirteenth century; static tilted arrangements survived for over two hundred years; antipodal dynamic schemes for about a hundred, and explicit disjointed dipoles for just under fifty. Each successive step possessed about half the longevity of its predecessor. Once again the quality and quantity of communicated observational evidence appears to have been the prime agent of change, more so than, for instance, the spread of hypotheses proper, through scientific networks or published treatises, pamphlets, and submitted longitude proposals.

This is particularly clear in the brief rise and fall of the disjointed dipole concept. In an increasingly desperate attempt to match theory with observations, the north and south magnetic poles were first placed at various fixed distances from the spin axis and each other. Soon they had to be disconnected altogether, in order to be able to pursue completely independent paths at different speeds and in opposing directions, while the number of parameters continued to grow. But eventually even this sacrifice proved insufficient to adequately explain and predict the Earth's intricate magnetic workings. To this day, the problem remains essentially unresolved.

Part II / In the Age of Sail

CHAPTER 5

Traversing the Trackless Oceans

Navigation involves the determination of position, course, and distance traveled during displacement. For most of the thousands of years of sail, it was an art and a craft rather than a science. In that terms are continually redefined to reflect changes in society, it may be more instructive to hear a contemporary explanation of the word than to consult a modern dictionary. Here is what William Bourne, a Gravesend gunner, had to say on the subject (in its practical sense) in 1574: "Nauigation is this, how to direct his course in the sea to any place assigned, and to consider in that direction what things may stande with him, and what things may stand against him, hauing consideration how to preserue the ship in all storms and chaunges of weather that may happen by the way, to bring the ship safe unto the port assigned, and in the shortest time."[1]

The safe, timely, and economical operation of the vessel was the responsibility of the master. On privately owned merchant vessels he also acted as the captain, who bore the general responsibility for the venture. On most ships of navies, large trading companies, and privateers, however, a separate individual performed the latter function. But even in the military environment, the position of masters was exceptional; although they tended to come from a lower social class than the commissioned officers, their pay roughly equaled that of lieutenants, and their special knowledge lent them both status and the power to overrule the captain in navigational matters. Naturally, the other officers were not ignorant of the issues involved; by the late seventeenth century, many naval lieutenants-to-be served for a time under a master, and captains started to undergo examination in general seamanship and navigation as well. But in disputes, the master's professional knowledge and expertise in guiding the ship safely to port counted heaviest. Fortunately, he could share this burden with two to four master's mates. Their capacity was twofold; the more experienced ones could substantially participate in the daily chores: standing watch, making observations, calculating position, and correcting charts and sailing directions. Meanwhile, the younger members of the team memorized the appearance of sea and coasts and learned the ropes: how to operate a bearing compass, judge

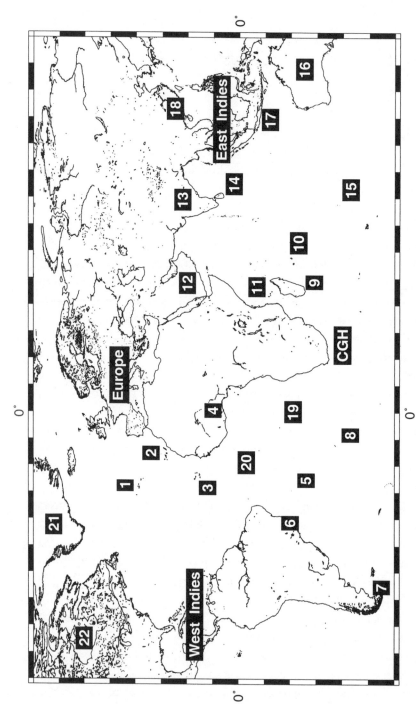

Fig. 5.1. Important way stations in the studied period: 1. Azores, 2. Canaries, 3. Cape Verdes, 4. Elmina, 5. Trinidade, 6. Rio de Janeiro, 7. Cape Horn, 8. Tristan da Cunha, 9. Madagascar, 10. Mauritius and Réunion, 11. Anjouan, 12. Arabia, 13. India, 14. Ceylon, 15. Amsterdam and St. Paul, 16. Australia, 17. Java, 18. Canton, 19. St. Helena, 20. Ascension, 21. Greenland, 22. Hudson's Bay. CGH = Cape of Good Hope, South Africa. Cylindrical-equidistant projection.

speed, leeway, and drift, mark the passing of time and tides, interpret the weather, sound with lead and line to determine depth and nature of the seabed, make celestial observations, and keep a daily reckoning (the so-called Day's Work).

In addition to the regular complement of navigation officers, local pilots could join the crew, either upon the approach to port, or during short journeys to destinations where the master lacked personal experience of the particular conditions and hazards. Although having been trained under a master, and thoroughly familiar with a particular stretch of coast, a pilot generally did not roam the vast ocean expanses unaided, which required additional skills. The relationship between the two types of navigators was not necessarily an easy one. On the one hand, the pilot was less versed in the intricacies of long-distance travel than the master. On the other hand, the master was out of his depth in alien coastal regions that were potentially full of treacherous shoals, shifting sandbanks, freak currents, and tides. Nevertheless, relinquishing control at the most dangerous stage of a voyage must have been hard at times, and clashes of competence were not uncommon. The case of the East Indiaman *Ascension* is telling, in that it ran aground off the Indian coast in 1609 after the inexperienced master had reputedly refused any aid from local pilots.

The dichotomy resulted not merely from a mincing of words regarding job descriptions; there was a fundamental difference between coastal and oceanic navigation in the age of sail. The former was based upon experience, compass, and lead, with the aid of printed and oral directions learned by rote, describing local tides, depths, and grounds. Sedimentary deposits varied widely in composition and were marked on coastal charts (as was depth). To bring up a sample from the seafloor, a weight on the end of a piece of line would have a hollow bottom filled with tallow, to which the grains would stick. Simultaneously, the procedure yielded the local depth. In principle, only the length of the line restricted the maximum to be gauged, but in practice it lay around a hundred fathoms (about 180 meters). This allowed access to most of the submerged parts of the continental shelves, far beyond the range of visible contact with land. It is a misconception to think that early mariners always hugged the coast, hopping from one reassuring landmark to the next. When one was in sight of land, one usually was also in shallow territory. The coast may have held the promise of trade and good living, but was to be won only after negotiating its natural defenses. Safety at sea lay in keeping some distance from shore. Even prehistoric sailors must have been perfectly able to spend many days on the open sea without losing their nerve.

Sea-lanes and Way Stations

Oceanic navigation differed substantially from coastal sea traffic, not by virtue of a larger distance from land, but because it required a more structured approach to determining position. Among the most important skills necessary to traverse an ocean were these: being able to convert observations of heavenly bodies into ship's latitude and local magnetic declination, carrying out a daily reckoning by calculation and chart, and keeping a proper logbook. Without frequent corrections based on soundings and land sightings, small errors could, after months of unchecked progress, take on disastrous proportions. Celestial observations could be obtained only under favorable atmospheric conditions; no instrument could accurately measure vessel speed or determine the effect of currents; longitude was sometimes merely a well-educated guess. It was therefore of vital importance to keep a thorough and extensive record of the ship's position and progress, as well as taking into account every possible clue that could contain helpful information to that effect.[2]

The energy necessary to traverse vast expanses of ocean came from two sources: winds and currents. Surface winds are driven by differences in atmospheric pressure and display complex patterns, with seasonal, short-term, and chaotic characteristics. A nonrotating Earth without continents would support a relatively simple system of atmospheric convection confined to certain latitudes, dependent on temperature. The introduction of large landmasses adds complications; not only is their distribution uneven over the globe, but their ability to absorb and retain heat differs markedly from that of water, affecting climate and winds. Naturally, local, regional, and global weather phenomena can also add disturbances.

Nevertheless, some large, regular traits can be identified. The most important are the wind belts, associated with certain latitudes. These follow the seasonal migration of the Sun with about a two-month lag, thus reaching their latitudinal extremes in February and August. The most famous is the westerly around forty degrees south latitude, dubbed "the roaring forties." The second seasonal system is that of the monsoons. Trade winds originating around thirty degrees latitude blow from both hemispheres toward, and alternatively across, the equator, meeting in a tropical convergence zone. As this zone shifts north and south during the year, due to temperature differences between sea and land, dry continental winds alternate with oceanic ones laden with moisture. Where their directions differ by 120 degrees or more, the term *monsoon* is applied.

The more troublesome regions from the perspective of sailing are those in between these giant conveyor belts. Areas enclosed by westerlies and trades tend to have variable, often unpredictable winds, whereas the battlefields of two opposing trade winds are the dead zones known as the doldrums. To most crews sailing from Europe to the Cape of Good Hope it was worth a detour past South America to avoid the doldrums in the equatorial Atlantic.[3] Another aspect of winds that complicates sailing is *leeway*. No matter how the sails are adjusted, unless a ship is traveling in the exact direction the wind is blowing, the vessel will experience sideways displacement. This effect is due to the wind force acting on hull and rigging, and has to be taken into account in dead-reckoning calculations. In addition to personal experience linking apparent and true course for all relative directions and wind strengths, the navigator could "shoot the wake," that is, take a compass bearing of the ship's turbulent trail relative to the ship's fore-and-aft line (see fig. 5.2).

Currents provided a second source of energy. Circular patterns turn clockwise in the northern hemisphere and counterclockwise in the southern; these large gyres could similarly work with or against a crew's intentions. The slave trade, for instance, could benefit from the Canaries Current to carry ships to equatorial Africa (to buy slaves), from the South Equatorial Current to bring them from Africa to the West Indies (to sell slaves and buy agricultural produce), whereas the Gulf Stream would take them home again in more northerly latitudes. In the South Atlantic, the Brazil Current flows down the South American coast, the Benguela Current up along African shores. Intended course and destination determined potential advantage or disadvantage; the Drake Passage Current could be a formidable opponent when trying to round Cape Horn to enter the Pacific, and the Agulhas Current could make northward travel between Madagascar and the African mainland nigh impossible. Another, more insidious effect of currents is *drift*, the whole mass of water surrounding the ship moving in a direction different from the steered heading. This displacement not only affected the ship's course but also was impossible to measure without a fixed reference.

Winds and currents could negate or enhance each other, or engage in variable interplay with the ship's course; out of port, a navigator's work is never

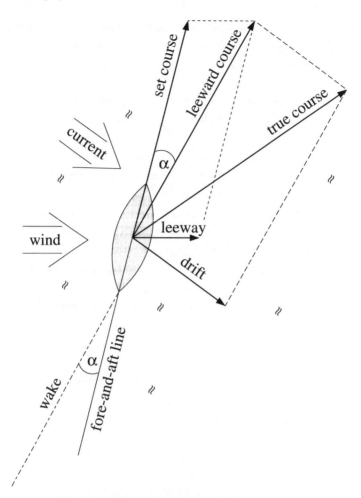

Fig. 5.2. Leeway and drift, and their effect on the course. The angle of the ship's wake with the fore-and-aft line is affected by leeway, but not by drift.

done. Wind and current systems thereby provided narrow corridors at certain places, for instance in the southbound crossing of the equator in the Atlantic Ocean. Too much to east, and the African doldrums would ensure arduous and slow progress along the coast, against the current and with variable winds. Too much to west, and the ship would end up on the "Wild Coast" of South America, forcing the mariners to return to the Azores in order to undertake a second attempt.[4] Seasonal weather variability and a conscious choice of route could furthermore cause directional bias to accumulate for weeks on end.

Take, for example, the voyage from the Cape of Good Hope to the East Indies, following westerly winds and currents for about a thousand miles due east, before heading north toward Java. In the absence of a reliable means to observe longitude, even a small structural underestimate of the ship's absolute speed could delay the decision to change direction sufficiently to cause the ship to run aground at the Houtman Abrolhos, off the Australian west coast.

This is one of the reasons why islands were of paramount importance. In the cited instance, it was the small twins of Amsterdam and St. Paul in the Indian Ocean which provided the much-needed positional check and correction. By staying on the latitude of these desolate, barren lumps of rock soon after leaving the Cape, seafarers found them easily and used them as signpost and milestone. Other islands had the advantage of high peaks, visible at great distances. Chains of volcanic isles frequently form long arcs, creating a target hundreds of miles wide, which would be very hard to pass without a positive identification.[5] In addition, natural harbors offered shelter for a badly leaking hull during prolonged storms, and some even supplied food and fresh water, or timber for ship maintenance.

The subject of food is closely tied to that of health, in particular the lack of sufficient vitamins. For much of the period under discussion, scurvy was an accepted part of life at sea, and yet another reason to make several stops along the way to distant destinations. Small islands could provide dietary supplements, rest, and recuperation; larger mainland ports had hospitals to cater to the sick and wounded. The journey by ship from Europe around Africa to the East Indies was simply too long to complete in one haul.[6] Although the average duration of such a voyage was about eight months, combinations of ill fortune and bad planning could extend it to over a year and a half. No craft was able to carry adequate provisions for such a long time, let alone keep them fresh in tropical conditions.

War and strained international relations could bar ships of one nation from seeking the safe havens held by opponents thousands of miles from the European political stage. Thus the Dutch, encroaching on the Portuguese spices monopoly in the early seventeenth century, could not benefit from Southeast African settlements in Portuguese hands, while sailing to the Indian subcontinent and beyond. Several failed attempts to capture these way stations forced them to find an alternative route, which happened to be a more efficient means to reaching the Indonesian archipelago. During the four Anglo-Dutch Wars, spice-laden Dutch East Indiamen tended to follow the more circuitous ap-

proach around Scotland rather than risk an encounter with Royal Navy cruis-
ers in the Channel. English captains then likewise steered clear of a hostile reception at Dutch Cape Town, resupplying at Madagascar instead. Contrastingly, in times of peace both harbors would be frequented, sometimes on the
same voyage. The French found refuge on Mauritius and Réunion, before proceeding to the Southeast Asian mainland. The English much preferred the
African corridor past Anjouan and Socotra and through the Arabian Sea. On
the other hand, smaller distances in North Atlantic waters enabled straight
runs from Europe to Hudson's Bay and North American shores. Trips to the
West Indies either proceeded directly from Europe, or via African ports and Atlantic islands, following the triangular current pattern described earlier. These
destinations were less of a navigational challenge in that *latitude sailing* (that
is, traveling to the latitude of the destination and then continuing dead east or
west until land was sighted) often sufficed for arriving where one wanted.

Charts and Logbooks

Regardless of destination, no competent navigator would venture far without proper charts. These can be divided into two categories. For coastal voyages of limited scope, *plane charts* were adequate. With degrees of latitude and
longitude of equal length, meridians appeared as parallel lines, instead of con-
verging toward the poles. The assumption of a flat Earth worked reasonably
well as long as the depicted area remained small, although the error did increase with distance from the equator. For oceanic travel plane charts were,
however, wholly unsuited, as plotted courses could no longer be approximated
by straight lines over large intervals. The problem was solved in 1569 by Flemish cartographer Gerard Mercator, and it was put on a firm mathematical footing thirty years later by Edward Wright. The *Mercator projection* was similar to
the plane chart in having straight meridians, but differed in the increasing distance between successive parallels of latitude nearer the poles, compensating
by vertical extension in exact proportion to the stretched horizontal dimension. The beauty of Mercator's invention lay in rhumbs becoming straight
lines; setting out a course across an ocean henceforth required only a simple
ruler. A price was paid for this advantage, though: surface areas and distances
became heavily distorted. But, a ship's heading being more easily determined
than its speed, the mariner's gain was evident.[7]

Logbooks testify that both plane and Mercator charts were taken along. The crew of the Royal Navy's *Tyger Prize*, upon departure from the West Indies for home in 1691, noted that from that day on longitude was "by the way of Mercator sailing"; the crew of the Dutch slaver *Prins Willem de Vijfde*, leaving Angola for Surinam (1759), logged a similar remark. Master G. Baker aboard the East India Company's *London*, upon reaching St. Helena from the Cape in 1761, determined his accumulated error of longitude by averaging the distance traveled on two plane charts and two in Mercator projection.[8]

It will come as no surprise that all charts were defective to an uncertain degree, in placing coastal features and whole islands at incorrect coordinates. Hydrographer F. Dassié had warned navigators in 1677: "One should not rely too much on charts, because they do not truly show in what place[s] the shoals and islands are,"[9] and he was right. As late as 1768, his colleague d'Après de Mannevillette even doubted the very existence of the islands of Martin Vaz, and in 1789 navigators on board the Dutch naval vessel *Ceres* found St. Helena to lie four and a half minutes more southerly than in their chart.[10] Part of the problem was that the information supplied to make and correct the charts was itself founded on dead-reckoned positions. Furthermore, differences between charts produced in neighboring countries can be attributed in part to the strategic value of hydrographical information. Seventeenth-century Dutch charts generally outclassed their foreign competitors, and were therefore smuggled to rivaling countries despite an embargo. Old Dutch copper plates sold for scrap had provided the source material for the first English maritime atlas (1671–89, by John Seller), and as late as 1769 Dutch East India Company instructions stipulated that sailing directions should be kept in a leaded chest while at sea, and dumped overboard in case of an unfriendly encounter. However, on numerous occasions, French logbooks identified charts as "Dutch chart" (*carte Hollandaise*), and many French mariners reckoned from the prime meridians used in Dutch charts (Tenerife and San Tiago), rather than the official French alternative (Ferro). Clearly, the protection of knowledge as envisaged by Dutch administrators fell far short of splendid isolation in actuality. By the late eighteenth century, the roles had been reversed, and French charts could occasionally be found on board Dutch ships.[11]

The complement of the chart was the logbook, the written record of systematic observation and calculation of direction, distance traveled, and position, as well as magnetic declination, wind, weather, currents, noteworthy landmarks and events, drafts of harbors, the appearance of coastlines, and other useful particulars. The accumulated experience also served as a reference for successive voyages by the author or other navigators, while hydrographers relied upon them to amend charts and sailing directions. Begun as a personal diary of maritime matters, the logbook evolved into a highly formalized data carrier with preprinted sheets, to be handed in for official inspection within a few days after the completion of a voyage.[12]

Explorer John Cabot was one of the first to lay down rules for keeping a logbook at sea, in the late fifteenth century. About a hundred years later, the navigation manual *Seaman's Secrets* by John Davis likewise stressed the importance of conscientious log-keeping. At that time, the practice was still left completely at the mariners' discretion. Little formalism existed in the way navigators processed the daily ship's progress: longhand entries could be one sentence short or continue for pages; quantities might be written out in words instead of Arabic numerals; longitude might be omitted altogether, or expressed in miles, or spherical degrees (so-called meridian distance).

The way a logbook stored information varied widely, depending on epoch, maritime organization, country of origin, and individual author. Free form started to give way to columnized layout in the 1620s on Dutch and English East Indiamen; in the 1680s, the Royal Navy, the French Navy, and the Compagnie des Indes followed suit. Another, block-shaped format with fixed designated areas for each type of observation first appeared in English logs of the 1660s. It was copied and modified by the Danes some sixty years later, but the Dutch and the French did not adopt it.[13] Companies and navies achieved increased standardization by supplying preprinted forms. A Dutch East India Company manuscript in the National Scottish Library seems to be the oldest surviving example (1640–43); the English East India Company had adopted the practice by 1705; and the French and the Danes in the 1710s and the 1730s respectively. The English "block" design became the official EIC standard in 1716; the Hudson's Bay Company adopted the same format from the 1750s. By the 1790s, even the merchant navies relinquished the written diary in favor of neat columns.[14]

What the daily entries should contain can be gleaned from examples in navigation manuals and official instructions. Of special interest here is the requirement to record magnetic declination. Some of the examples given in textbooks lacked a column for the *variation of the compass*, as it was generally called at sea, while certain directives stipulated that all opportunities to observe it should be seized.[15] Early logs often feature the geomagnetic field in the remarks, while later ones have a single space or column reserved per page.

Ideally, the text would clearly distinguish the observed magnetic declination and the value allowed for in steering, dead-reckoning, and shooting land-marks. The two are not necessarily related, as will be explored in the final chapter. Other distinctions of interest concern the type of observation made, as regards the time of day, *amplitude* or *azimuth* (that is, on the horizon or above it), and, in rare instances, with what make of compass the measurement was performed. The Dutch employed a system of over thirty abbreviations, whereas certain Danish and French logs had hourly tables for each day to designate the time of sighting, and applied simple references in other instances. Observational practice figures in the next chapter.[16]

Latitude and Longitude

On 28 September 1690, the English East India Company's ship *Chandos* was in dire straits, its captain John Bonnell lamenting: "Not having an observ. at noon nor any these three dayes we durst not run but lay by till [it will] please Providence to give us an oppertunity to observe." The ship continued aimlessly adrift in the South Atlantic for a week, until finally, on 5 October, "toward noon the sunn appeared frequently so as the officers made a shift to gett an observ. . . . I could not observe myself, being ill at present of the gout, w.ch hat afflicted me these ten days past . . . They advised me and pressed me to make saile, being as they say assured their obser. was good; supposeing it to be so I must allow the ship to have driven to the N.ward considerably."[17]

The most important, and potentially most accurate, observation on board was that of latitude. The distance from the equator as derived from the altitude of various heavenly bodies formed one half of the coordinate pair defining position on Earth. If often measured, it provided a powerful constraint on dead-reckoning estimates. When overcast skies prevented such a correction for many days, the consequences of accumulated error could be grave. Captain Bonnell's decision to lay by may have been lacking in bold adventurous spirit, but was nevertheless perfectly justifiable as a means to minimize the danger to cargo and crew.

Determining latitude had been a long-established practice by land-based astronomers, and it was initially their instruments, the quadrant and astrolabe, that became adapted for use at sea. A more practical device on deck was the cross-staff, a stick with separate transoms of different dimensions sliding along its length. Aligning the lower end of a crosspiece with the horizon and the top

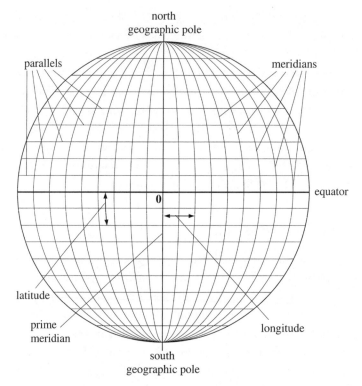

Fig. 5.3. Arc measures to designate position on Earth. Latitude designates the distance (north or south) from the equator (from 0 to ± 90°); all points on the same latitude together form a parallel of latitude; longitude denotes the distance (east or west) from a chosen prime meridian (from 0 to 360°, or from 0 to ± 180°); all points on the same longitude together form a meridian of longitude.

with the object to be sighted, the latter's altitude above the horizon could be read off on a graduated scale. Unfortunately, the observer was unable to peer along both ends of the transom at the same time. An additional problem was having to look directly into the Sun. The introduction of the backstaff, invented by John Davis in 1594, therefore constituted a considerable improvement. The next major stride forward was the octant, developed around 1730 by John Hadley in England, and independently by Thomas Godfrey in Pennsylvania. A moveable and a fixed mirror effectively doubled the scope of the 45-degree-arc frame to bring celestial object and horizon into apparent coincidence. The same principle underlay the sextant, invented some thirty years later. The latter device had a larger vertical range (120 degrees) than the octant

(90 degrees). Whereas the cross-staff had offered a maximum accuracy of about half a degree, these eighteenth-century improvements increased it to mere tens of minutes of arc. As numerous textbooks stressed, observation of latitude was one of the basic necessities of oceanic navigation.[18]

Latitude equals the observed altitude of the visible celestial pole, in the zenith at the geographical pole and at the horizon on the equator. In the northern hemisphere, the Polestar (*Polaris*) resides very near to this point. At present the distance is less than one degree, but five hundred years ago the difference amounted to more than three, due to the slow precession of the Earth's axis. Given this distance, Polaris seemed to complete one tiny revolution around the celestial pole in the space of about twenty-four hours. Late in the fifteenth century, the relative positions of neighboring stars during this daily revolution became codified in a table. Sometimes referred to as the *Regiment of the North Star*, it listed the necessary adjustments to convert the star's altitude to that of the true celestial pole.[19] In the southern hemisphere, the bright stars of the "Southern Cross" constellation served in this capacity, the Polestar being invisible there on account of the intervening body of the Earth.

The only stellar object making regular appearances in both celestial hemispheres is the Sun. Perennial ferryman between the tropics of Cancer and Capricorn, its position at any time during the day is not only dependent upon the observer's latitude, but also on the date in a four-year cycle. Spherical trigonometry and plenty of patience reduced these movements to a set of Sun's declination tables, yielding celestial coordinates for all 1,461 days. European astronomers have calculated and produced several such tabular predictions from the thirteenth century onward.[20] In the maritime realm, almanacs eventually combined these *ephemerides* with instructive examples and information on fixed stars, lunar phases and movements, tides, and the longitude of ports. In the course of the seventeenth century, these publications also started to include sailing directions and logarithmic tables of trigonometrical functions. Initially, the accuracy of the stated coordinates left something to be desired. The Greenwich Royal Observatory, built in 1675, was therefore intended to provide navigators with more precise astronomical data, in particular regarding the Moon. By 1766 Astronomer Royal Nevil Maskelyne supervised the observatory's publication of the official English *Nautical Almanac*. Meanwhile, Jean Picart and other astronomers of the Parisian Académie des Sciences started issuing their own *Connaissance des Temps* in France from 1679, a task eventually transferred to the Bureau des Longitudes upon its foundation in 1795. The Dutch had to wait until 1788 before their "Committee for the Determination of Longitude and Improvement of Charts" (1787–1850) supported their own version, based upon the English example.

Maskelyne's main reason for devoting much time and energy overseeing the publication of the nautical almanac was his desire to solve the standing problem of determining longitude at sea, the other half of the coordinate pair. He adhered to the *lunar-distance method,* which derived local time and meridian from the Moon's rapidly changing position relative to the fixed stars. Originally propounded in theoretical form around the turn of the sixteenth century, it received a new impetus in the 1750s, when Göttingen professor Tobias Mayer compiled extensive tables of the Moon's complex wanderings. These relied upon new mathematics developed by Leonhard Euler, and were the first with enough precision to allow practical application; limitations in instrumental accuracy and astronomical theory had frustrated attempts made before that time. Published in 1752, and submitted to the English Board of Longitude three years later, Mayer's tables offered Maskelyne an excellent foundation for constructing a table of the Moon's position for every noon and midnight. The sextant, with its ability to measure angles larger than ninety degrees, was the principal tool in the table's use at sea.[21]

The underlying idea of any celestial timepiece, be it the Moon serving as clock hand or, for instance, the eclipses of Jupiter's four largest satellites, was that the predicted time of the event as observed on a standard meridian could be compared with local time. Each four minutes of difference then equaled one degree of longitude. Yet another approach was through mechanical solution, by inventing a chronometer able to withstand all disturbances impeding its regularity. As in the lunar case, theory preceded practice by over two hundred years. Ships' movements and changes in temperature, humidity, and (latitude-dependent) gravity were finally defeated by the fourth prototype of instrument maker John Harrison, tested at sea in the early 1760s.[22]

On 9 February 1765, the English Board of Longitude formally approved both Harrison's timekeeper and Mayer's lunar tables. Unfortunately, it would take several more decades before these two solutions became part of standard shipboard practice. For lunar distances, the required extensive calculations detracted from its appeal to sailors. By comparison, the chronometer yielded almost instant results, but its complicated construction made the instrument prohibitively expensive and initially unfit for mass production. Complaints about limited issue and supply continued far into the nineteenth century. For future reference, it is important to note that from the 1760s onward, observed longitude started to make timid appearances on the navigational stage, but failed to make a substantial impact until the 1790s. Only then were logbook formats revised to accommodate the chronometer's results, and was the mariner's fundamental uncertainty as regards easting and westing gradually lifted.[23]

Until that time, masters and mates had to rely on other sources to improve their longitudinal estimates. Corrections by means of identifiable land sightings and observation of latitude have already passed review. A group of ships sailing in convoy could further lessen individual error by taking the combined average of measurements made on all vessels. Lone ships often welcomed the opportunity to compare notes, as happened in 1752 when the Royal Navy's *Tryall*, on its way home from Newfoundland, met an outbound ship from Bristol near the Scilly Isles. The craft that had recently left port had had far less opportunity to accumulate positional error, and thus its fix was happily copied in the naval logbook.[24]

Many are the tales of gross misjudgment in reckoning, ranging from five to ten degrees in longitude. Samuel Dunn in 1775 assumed it could amount "from one to two or three hundred miles or more."[25] But appearances can be deceptive. The same psychological trap is sprung with shipwreck stories. A taste for the unique and sensational can bias interpretation at the cost of more mundane statistics obtained from a representative sample. Spectacular mishaps leave much more of a lasting imprint than dull facts like the following: over a period of 193 years the Dutch East India Company suffered some form of shipwreck on a tiny 3 percent of its 8,190 ocean crossings to and from the East. Admittedly, several instances of appalling mistakes in reckoning did occur, but these are the exceptions. The above is not meant to belittle the gravity of the problem, nor the skills of the navigators doing battle with it. Among the causes already discussed are the difficulty of estimating leeway, the set of currents, and the inaccuracy of tables and charts. Related to the latter is the metrological matter of the actual length of a mile and a degree.[26]

Vessel speed could be roughly established by tracking the time it took a floating object or a patch of foam to pass a given length of hull. A more sophisticated approach used a *logline*, a float attached to a light piece of rope with knots at fixed intervals. The former being dropped from the stern, the knots would pass through the hands of the person paying out the line during a fixed interval, measured by sandglass. Each counted knot assumedly corresponded to one nautical mile per hour. This left plenty of scope for inaccuracy, due to incorrect spacing between the knots, dragging by insufficient line, turbulence in the water, or the instruments not performing properly, as was discovered aboard the English East Indiaman *Neptune* in 1766, when the half-minute sandglass turned out to be five seconds short.

Despite the multiplicity of factors involved, the stock reply for many discovered discrepancies between presumed and true position was to blame the currents, the most elusive of all intangibles involved.[27] Tides could also be at work, such as the indraft of St. George's Channel, the flood supposedly setting ships bound for the Channel from the west so far northward as to force them to run up the Bristol Channel or the Severn Sea instead. In an anonymous "Advertisement Necessary for All Navigators Bound up the Channel of England" (1701), astronomer Edmond Halley pointed the finger at a different culprit, namely incorrect allowance being made for local magnetic declination. Steered courses are reliable only if the recorded bearings are true; any under- or overestimate of the local difference between magnetic and geographic north will lead to an erroneous positional estimate.[28] Therefore, compensation had to be constantly adjusted, on the basis of measurements, sailing directions, and experience. The multifarious ways in which navigators dealt with the encountered peculiarities of Earth's changing magnetic field once constituted an important part of oceanic navigation, as the next three chapters will set out to explore.

CHAPTER 6

Following the Iron Arrow

The invention of the compass has been placed on a par with that of gunpowder and the printing press, grouped among the most fundamental technological advances in early-modern times, and hailed as a paragon of human progress. Whether or not these laurels are justly deserved is debatable, but few instruments combine a similar simplicity of design with a comparable impact on society. The ability to determine direction magnetically allowed the traveler to set a course and keep it, the surveyor to map territory and landmarks, the miner to locate bodies of iron ore, the navigator to reckon his position, the hydrographer to chart the sea-lanes, and the natural philosopher to investigate the Earth's magnetic field.

Widely available iron and lodestone resources across the world offered many cultures the chance to discover these various applications of the instrument. At sea, the needle initially floated in a bowl of water and needed frequent remagnetization. In 1269, the French engineer de Maricourt described the first dry amplitude compass, which had the needle balancing on a pivot inside a box with a graduated scale and a sighting rule. Its purpose was to take bearings of objects near the horizon. The next step forward was the fourteenth-century addition of a card (or *fly*), a stiffened paper disc depicting the thirty-two-point rose of the winds, to which the needle was affixed underneath. This combination offered the advantages of simultaneous identification of all directions and a larger moment of inertia. Since the fifteenth century, the four cardinal points traditionally bore blue, and the half cardinals red. The card's north frequently also carried an ornamental lily (fleur-de-lis).

Mariners in the early sixteenth century could determine the locally variable magnetic quantity with the aid of a *gnomonic compass*; they employed the shadow cast by its central upright style at equal times before and after noon to find the local meridian, to which magnetic north was compared. This hand-held device was eventually succeeded by the more familiar round or square box made of wood, copper, or brass. This "compass bowl" had a wind- and water-tight glass lid and was suspended in gimbals. The needle was initially made of

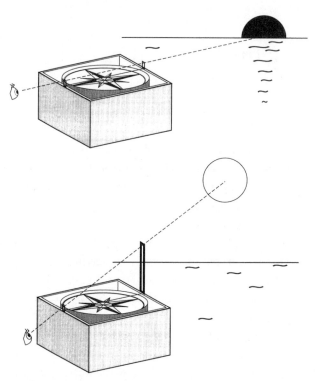

Fig. 6.1. The main difference between an amplitude compass (*top*) and an azimuth compass (*bottom*) is the shape of the visors; the former's visors (short, equal length) allowed taking bearings of objects only very close to, or on the horizon, whereas the latter's visors (tall, unequal length) offered a much improved range in sighting altitude. The distinction in terms follows Continental practice.

iron wire, bent in the shape of a lozenge. But by the second quarter of the eighteenth century, it came to be replaced by a single, straight piece of steel, which could be magnetized more strongly, and for a longer time. Weak magnetic force had earlier been one of the main compass defects, next to improper card orientation, and friction and wear on pivot and cap.

One has to distinguish clearly between a *steering compass* and a *bearing compass*. Two steering compasses stood in separate compartments of the binnacle in front of the helm, in order to maintain a set course. Their cards usually carried only a graduation in points and parts thereof, in the familiar star-shaped pattern. Some also contained a compensation mechanism for local declination, as will be discussed later. A bearing compass, on the other hand (also known as *observation compass,* or *variation compass*), had a card graduated in degrees, and additionally featured a mounted, sometimes rotatable sighting apparatus on top of the bowl, in order to "shoot" celestial bodies, landmarks, or the ship's wake. These compasses had no fixed place on deck; navigators could momentarily set them up on a tripod or stool as occasion demanded. In its early form, known as an *amplitude compass,* the sighting mechanism consisted of two small visors of equal length on opposite sides of the box, an arrangement that only allowed taking bearings in the horizontal plane, directly over the card.

The 1750s, however, saw the swift rise in English, Danish, and Dutch navigational practice of the *azimuth compass.* This innovation had taller visors of unequal height, which allowed the measuring of objects at a far greater altitude above the horizon than had previous models. Remarkably, French eighteenth-century improvements to the instrument did not cater to this new type of observation. This failure was due to both conservatism by maritime practitioners and institutional lethargy on the part of the French Navy, which processed most of the inventions submitted. Existing imperfections in English compasses eventually led to the adoption of Ralph Walker's design. It bore resemblance to a sundial, and magnetic declination could instantly be read off. It became the Royal Navy's standard issue by the late 1790s, after a long battle for recognition, replacing an earlier model by "magnetician" Gowin Knight.

Observation compasses served to keep track of magnetic declination, which changed from place to place during an ocean voyage. Measurements and past experience then provided an estimate of how much the steering compass had to be compensated in order to keep a true course. This so-called compass allowance subsequently found its way into the dead-reckoning calculations and the logbook. Surviving manuscripts offer a revealing glimpse into this maritime world; this chapter intends to address various aspects of geomagnetic navigational practice, based on the largest number of relevant sources ever examined for this purpose.

Awareness

Among the rich holdings of the Bodleian Library in Oxford is a text in the Tanner Papers "Concerning Navigation" from the 1640s, which contains the following warning regarding the compass:

If it be not considerately handled, it may occasion great errours. For by experience it is found that almost in all meridians there is some variation of it, so that the needle ... points either easterly or westerly, and that with such an uncerteinty, that in some meridians this variation is many degrees, in some a few, in some it is hardly sensible. If therefore especially in long voyages seamen shall either neglect, or contemne this property of the compasse, besides the loosing of much time in performing of the voyage, they may runne into many other inconveniencies.[1]

To what extent was this counsel heeded by fifteenth- and sixteenth-century mariners? Assuming that navigators had a vested interest in their own safety, awareness of the problem would elicit some form of response. The principle of compensation was straightforward: should the magnetized needle point ten degrees east of true north, then all steered headings and bearings had to be shifted ten degrees to the west to regain true directions. If the value to be corrected for was variable with position, then the allowance needed commensurate adjustment on a continual basis. Such corrections affected daily dead-reckoning calculations, and have therefore left an unmistakable trace in navigational logbooks. Unfortunately, very few such manuscripts from the fifteenth and sixteenth centuries have survived. In addition, as long as the continental shelf was not left, reliance on the compass was limited, since plenty of opportunity was usually available to correct estimates through other means. Ocean travel started in earnest in the sixteenth century, but at that time not everyone considered magnetic declination a real phenomenon. Wide acceptance was hampered by certain highly respected navigators, who initially rejected the whole notion outright. Only through the gradual amassing and dispersal of data did the concept win growing support, until even the staunchest skeptics could no longer deny its existence.

Christopher Columbus must have been one of the first to sail from an area of *northeasting* (Europe) across an area of very low declination (around the Azores) to a region of *northwesting* (America). Using the Polestar as a reference, he made three observations in September 1492. Regrettably, the original logbook of the voyage is lost; all information is based on secondhand accounts. The earliest sixteenth-century source is an anonymous Portuguese work from 1502, *Of the Needle* (*Das Agulhas*), which mentions how navigators, upon entering the Indian Ocean, verified that their needles registered no declination at Cabo das Agulhas (the "Needle Cape," South Africa). This remark could have referred to any of the contemporary Portuguese fleets of exploration (Dias

1488, da Gama 1497, Cabral 1500, or de Nova 1501). Other sources include a journal from 1505 written by Francisco de Almeide of a voyage from Lisbon to India, recording needle deflection en route from Brazil to the Cape. João de Lisboa's treatise on the nautical needle (in manuscript in 1508, published in 1514) furthermore contains a passage on determining declination in the southern hemisphere, sighting a star of the "Southern Cross" constellation. A treatise of Jean Rotz (1542) and a later work by William Cunningham (1559) moreover gave magnetic readings obtained at Dieppe, Corvo, Newfoundland, and the Brazilian coast.[2]

Nor should one underestimate the influence of cosmographers as agents of knowledge dispersion. The work of the Faleiro brothers in Spain is a case in point. Their nautical manual contained extensive instructions on how to actually observe the magnetic quantity. Even while still in manuscript form, it traveled the world during Magellan's circumnavigation (1519–22); about a decade later it finally appeared in print. Another example is the 1536 conference of cosmographers and masters at Seville's "Trade House" (Casa de Contratación). Held to draft a sailing direction for the West Indies, the meeting involved substantial arguments concerning local values of declination. Meanwhile in Portugal, Pedro Nunes (professor of mathematics and royal cosmographer) instructed explorer João de Castro in the fine art of magnetic observation, prior to the latter's 1538–41 exploration of the East Indies. This resulted in forty-three high-quality observations, which soon found their way into published roteiros (sailing directions). In the 1570s and 1580s, other famous Portuguese navigators engaged in similar activities, including Vicente Rodrigues, Dieggo Affonso, and Aleixo da Motta.[3]

But not all experts were convinced that magnetic declination was not simply due to error. Probably the most widely read author denying the existence of straying needles was Pedro de Medina, astronomer, geographer, cosmographer, teacher, and examiner of navigators. His arguments for rejection stem purely from theoretical reasoning, and are valid within their own framework. His 1545 *Arte de Navegar* sums up three main reasons:

1. Needles made from the same steel and magnetized by the same lodestone do not behave differently, so if one is taken west of the Azores and the other east, why should they act dissimilarly? Furthermore, the [geographical] pole is an imaginary fixed point, which does not wander about.
2. Needles do not physically change depending on their location.

3. If the needle were to truly err, it would be following another [kind of] meridian, which would imply the existence of an infinite number of poles, which cannot be.

Attributing the cause to the needles rather than to forces acting upon them lies at the root of the first and second misconception; a failure to distinguish between geographical and magnetic poles is responsible for the third. An assumption of a dipole system (as opposed to a more complex field pattern) underlies all three as well. Another notable remark figures in the fourth chapter of de Medina's work, listing the perceived inconveniences that would accompany magnetic declination: all charts would be riddled with orientation errors, and dead-reckoning would be flawed. "Those adhering to this baseless notion therefore should be alert," warned de Medina, as if the people struggling to deal with the phenomenon would bring upon themselves the evils they were compensating for.[4]

De Medina's text was not soon forgotten. As late as 1587, his three original arguments still echoed in a text by Diego Garcia de Palacio. Even more remarkably, seasoned navigator Pedro Sarmiento claimed that only people with little navigational experience believed in magnetic deflection from true north, and any variability outside the normal movement range of half a compass point (5°38′) could be cured by simply keeping the needle well oiled. But the reactions these opinions provoked were not mild; among de Medina's critics were Martin Cortés (1551), Toussaints de Bessard (1574: "De Medina has not understood the variation of the magnetized needle"), Michiel Coignet (1581), Robert Norman (1581: "It appeareth that he had no more regard to the variation then many mariners in these dayes"), William Borough (1596: "Perceiuing the difficultye of the thing, and that if they had dealt therewith, it would haue utterlie ouerwhelmed their former plaucible conceites"), Robert Hues (1592), Johannes Kepler (1598), Henry Gellibrand (1635), Edward Wright (1657), Abraham de Graaf (1658), Guillaume Denys (1666: "Navigators of his time were much less intelligent than those in the present"), Ioannus Riccioli (1672), Claude Millet Dechales (1677), and Pieter van Musschenbroeck (1729).[5] Although de Medina has succeeded in being remembered by posterity, it seems doubtful whether he would have appreciated the manner in which this honor was bestowed. By the advent of the seventeenth century, no reputable author in the field dared to question the existence of magnetic declination and its variability with position.

Another aspect of geomagnetism is its time-dependent change, as published by Gellibrand in 1635. The concept was quickly accepted by a wide audience, perhaps because variability over time was able to reconcile undated, seemingly conflicting observations made at the same place, but at different times in the past. Acknowledgments of secular variation in seventeenth-century logs and sailing directions are more difficult to find. A 1669 description of the Magellan Straits by captain John Wood recalled declination to have changed eleven degrees westerly in London since 1591, while in Port Desire (Argentina) an equal easterly shift had been observed. The French *Routier des Indes Orientales et Occidentales* (a published sailing direction from 1677) remarked on magnetic values generally tending with time to diminish in easterly direction while increasing westerly. Similarly, the master of the *Prospect* in 1682, after inspecting his compass and comparing the measured value at Galle (Sri Lanka) with the one in a 1675 logbook, had to conclude that he could not attribute this "error" to anything else but the declination's change over time.[6]

Types of Observation

A sample of sixty-three navigation manuals, straddling the interval 1574–1787, serves to illustrate the many methods employed to obtain information about local magnetic north. The rationale is the same for all: the position(s) of a celestial object at a certain time yielded the direction of geographic north, to which the needle's stance could be compared; the difference equaled the magnetic declination. Figures 6.2–6.4 depict the principle as applied in several cases. When a star reaches its highest point in the sky, it is standing in the meridional plane, and the latter's angle with compass north equals the local declination. The Polestar is special in that its position roughly coincides with the celestial pole, the imaginary center to which the Earth's rotation axis is pointing. It therefore never strays far from the meridian line. Equally straightforward is a *double amplitude* observation. The local meridian equals half the angle between a heavenly body's rise and set, to which the needle's orientation could once again be directly compared. The land-based method of *equal altitudes* was closely related, the only difference being that the object was shot twice at the same level above the horizon, instead of directly upon it. This required a second observer to record its altitude, so the correct height of the second sighting could be established.

A slightly more complex measurement was *single amplitude*; this involved

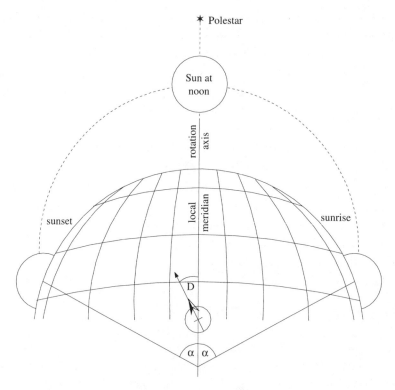

Fig. 6.2. Polestar, noon, and double-amplitude observation. The culmination of Sun and Polestar immediately yield the local meridian, to which local magnetic north can be compared; alternatively, the meridian can be found by taking the middle of the angle (2α) between sunrise and sunset.

taking the difference of the Sun's true and magnetic direction on the horizon relative to east or west. The true orientation (so-called amplitude) was stated in solar declination tables, which predicted its bearing for every day of a four-year cycle for most latitudes. Such tables featured in all nautical almanacs and many textbooks, but were also sold separately. Alternatively, navigational enthusiasts could generate these data on the spur of the moment by spherical trigonometry. Logbooks, however, seldom bear witness to these extensive calculations, suggesting that mariners normally relied on a table.

The most difficult observation was *azimuth*. The term denotes the horizontal arc between the local meridian and the vertical plane through the observer and the sighted object. The technique involved shooting the Sun (or any other star) at any substantial height above the horizon, requiring one observer to

Table 6.1 Methods of Measuring Magnetic Declination, Grouped by Origin

Origin	Method	Altitude	Table	Calculation
Land:	Land			
	Culmination	Yes		
	Polestar		Yes	
	Equal Altitude	Yes		
Sea:	Double Amplitude			
	Single Amplitude		Yes	Yes
	Azimuth	Yes	Yes	Yes

Note: Land = land-based astronomers and surveyors; Sea = navigators at sea.

measure altitude while the other sighted the object through the extended compass visors of the azimuth compass. As in the case of the Sun's noon measurement, a wire strung horizontally or diagonally over the card could cast a shadow instead, if the Sun stood high enough. This latter method had the advantages of requiring neither particularly tall visors nor having to look directly into the glare, but did involve more mathematical handiwork.

Table 6.1 lists the various main methods of observations, subdivided into two groups. The first contains the oldest types, which were imported into the maritime world from land-based astronomers and surveyors. The second group consists of amplitude and azimuth techniques, and represents the overwhelming majority of measurements found in ships' logs. The "Altitude" column denotes whether a simultaneous observation of height above the horizon was necessary, while "Table" marks the need for tabulated predicted celestial positions. The last column indicates whether the value could be immediately compared with the needle, or required subsequent calculations beforehand.

Observations made on land are the most venerable tradition, dating back to a time when keeping the apparatus level on deck was still a problem. Masters in the keep of Spanish explorer Pedro Fernandez de Quiros took refuge to it during their late-sixteenth-century exploits in the Pacific, as did Dutch navigator Philips Grimmaert while reconnoitering Indonesian coasts in 1599. Other examples include the captain of the *Pentecost* off North American shores in 1605, George Downton at the Cape of Good Hope (1614), and William Baffin in Hudson's Bay a year later; all took the trouble of setting up their equipment on dry land. Rodrigo Zamorano's navigation textbook (1581 and later editions) gave detailed instructions on how to trace a meridian line with the aid of a gnomon placed at the center of a circle. The two moments when the Sun's shadow precisely reached the rim constituted an equal-altitude reading on land.[7] With the advent of better gimbals supporting the compass bowl, the practice seems to have ceased altogether. Eighteenth-century logbooks hardly ever make mention of it, and the references in navigation textbooks by Bouguer (1760) and Dulague (1775) appear to be merely added in the interest of completeness rather than as a serious suggestion. By this time, navigators must be deemed perfectly able to make observations from the ship's deck.

Celestial objects reaching their culmination provided a second source of direct information regarding the meridian line's orientation. Whether it was the Polestar doing its rounds, bright stars halfway along their heavenly traverse, or the noonday Sun, their meridional passage could be directly translated into a value of declination. In the case of Polaris and the Sun, determining the body's altitude at the same time offered the boon of observed latitude. The frequency with which certain methods appear in navigational textbooks suggests that shooting the Polestar was mainly an English seventeenth-century pastime, while watching the culmination of the Sun and other stars seems to have been primarily a French privilege throughout the investigated period. The same holds for the method of equal altitudes, featured in quite a number of (predominantly French) textbooks, but little practiced.

Types of the second category make up the majority of all observations, both in navigation manuals and logbooks. Of all textbooks, the Dutch are most heavily biased in their favor. In most instances, the object of study was the Sun, but French manuals exhibited a peculiar predilection for observing other stars. Remarkably, this choice is poorly reflected in their logbooks. This group of methods is also predominant in worked exercises and related rules-of-thumb, found at the back of some logs.[8] As table 6.1 shows, the simplicity of taking double amplitude over other types is clear (no altitude or declination tables were necessary, yielding a direct result without any calculations). Its main drawback was its dependence on two consecutive clear sightings. These usually were sunset and sunrise (in that order, since the nautical day started at noon).[9] Atmospheric conditions frequently being less than optimal, the single amplitude proved a worthy alternative. Even though it required a solar declination table and some calculation, it surpassed all other types of measurement in frequency of use during the whole of the seventeenth, and part of the eighteenth century.

In order to establish whether preference for either sunrise or sunset existed, a sample of 899 logbooks was examined that had both kinds of observation. The outcome was a rather unequal division in favor of evening amplitude

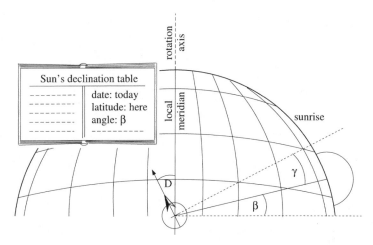

Fig. 6.3. Single amplitude observation. The Sun's apparent position (γ, north or south of east [at sunrise, depicted here], or west [at sunset]), as determined with the compass, is compared to a table containing the predicted true value (β, so-called amplitude) for the observer's latitude at the appropriate date; their difference equals local magnetic declination.

across the line. The uniform preponderance of observations after noon was due to a purely practical consideration, namely that it gave the observer more time to prepare. The descending Sun served as an obvious marker for the amount of time left before it would disappear from sight, whereas shortly before dawn, working in twilight, it was difficult to predict the exact time of emergence above the horizon. This is due to the fact that the Sun's apparent vertical speed is dependent on latitude and its angle of rising, which varies with latitude and over time.

There are, however, three problems associated with the single amplitude observation. The first is encapsulated in the word *single;* for better or for worse, one could at most take only two amplitude readings a day. The second was the dependence on horizon sighting conditions at those two instances; tropical haze or low clouds sometimes inhibited proper measurement. The third involved the incurred error through atmospheric refraction, which is largest at the horizon. The solution lay in sighting a celestial object somewhat higher in the sky and taking into account the horizontal displacement relative to its point of coincidence with the horizon. Azimuth observations were therefore not restricted to any particular time, but only by the sighting facilities of the compass.[10]

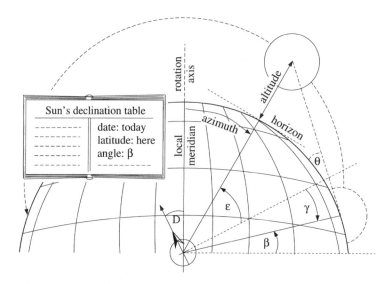

Fig. 6.4. Azimuth observation. The Sun's bearing, altitude, and angle of rising (γ) at a certain date and latitude determine its position at sunrise (γ, north or south of east as determined with the compass); this position is compared to a table containing the predicted true value (β) for the observer's latitude at the appropriate date; their difference equals local magnetic declination.

The technological development of azimuth visors thus made it possible to take readings at a greater altitude. The measurement itself required the combined efforts of two or three people: one for keeping the compass aligned, another to read off the compass dial, and a third to take the object's altitude with an octant or similar device. Due to the constant movement of the ship's deck, the first observer had to shout when orientation was momentarily perfect. The sighted object was most often the Sun, but other stars could serve equally well. Of course, the increase in the number of observers and measurements widened the scope for inaccuracies. The ability to take a combined average of several readings shortly after one another fortunately helped to decrease the observational error. Naval lieutenant Edward Harrison was ahead of his time when he advised the mariner in 1696 to "trust not to one observation, when you can have the medium of 5 or 6 or more, nor to one amplitude when you may have the mean of 3 or 4 azimuths."[11]

Initially, the amplitude observation compass (which did not carry tall visors) provided incidental azimuth sightings of up to about ten degrees above the horizon; the advent of special azimuth compasses in the eighteenth century

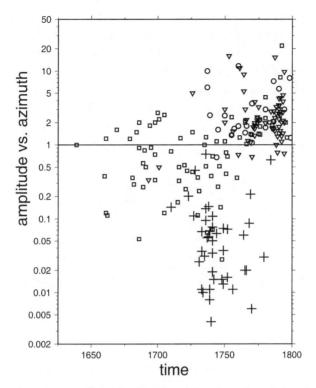

Fig. 6.5. Scattergram of amplitude versus azimuth on board East Indiamen, over time (based on logbooks featuring both types). *Square*, English; *cross*, French; *inverted triangle*, Dutch; *circle*, Danish. The ratio of azimuth divided by amplitude is plotted on a logarithmic scale; level 1 denotes equal amounts of observations of each type; above it, azimuth dominates; below it, amplitude is predominant; all navigators except the French show increasing use of azimuth over time. $N = 207$.

heralded an era of much wider application. Nevertheless, the transformation in observation practice from amplitude to azimuth was gradual, and differed across national borders. The acceptance of azimuth as a superior means of obtaining local magnetic declination was as much reliant on adequate technology as on the implementation of the new, longer calculations and awareness of the flaws inherent in single amplitude measurements. Astronomer Edmond Halley was still following conventional practice on his Atlantic magnetic survey in 1699, taking azimuth only as a last resort: "Finding it scarce possible to gett an amplitude in this cloudy and foggy climate, I am forced to take the Sunn's azimuth when he is low."[12]

The first half of the eighteenth century, however, brought a change of mind-set for all navigators except the French (see fig. 6.5). Both English and Dutch navigators experimented with the azimuth concept long before it became the norm. For East India companies in both countries as well as the Royal Navy, the 1750s was the decade when the conscious shift toward azimuth took place. The Dutch Admiralties seem to have followed a decade later. The Danish record is rather remarkable in that azimuth was predominant from the start of their oceanic endeavors in the 1730s. This holds both for their Asia Company and for their navy. Perhaps their relatively late start has enabled them to profit from novel technology without the slowing effect of conservative practice. As for the French, only a small fraction of sampled logbooks can boast of azimuth at all, and when ordered chronologically, neither the Compagnie des Indes nor the French Navy displayed a decisive commitment to it at any time during the eighteenth century.[13] This is all the more noteworthy since most navigation texts discussing azimuth observations were French. Once again, textbook theory did not necessarily concur with logbook practice.

Frequency and Accuracy

The quantified processing of magnetic measurements from historical log-books offers a chance to investigate various aspects of observational practice. Once the data is in machine-readable form, the material can answer a host of questions, even years after the actual sources were read. How often were observations made per day? How many were taken during a whole voyage? Did notable differences exist between countries or sailing routes? How high was the accuracy rate that was striven for and attained? Which factors contributed to instrumental error? Were navigators aware of these, and did they compare compass performance? The following paragraphs will briefly address some of these issues.

Observation frequencies can be separated into three types: the number of readings taken in short succession to obtain a single estimate at a certain time, the number of such measurement sessions during a nautical day, and the total number taken over a whole voyage. Unfortunately, the logbook often does not specify whether a recorded value represented one sighting or one session of multiple sightings. A purely qualitative remark in this respect concerns the difference between practice on regular runs to known quarters, and voyages of discovery where conscientious observation sometimes bordered on the fanat-

Table 6.2 Number of Logbooks and Average Number of Observations of Magnetic Declination (avg. N) per Voyage on Ships from Different Countries, Along Several Routes

Route	English		French		Dutch		Danish	
	logs	avg. N	logs	avg. N	logs	avg. N	logs	avg. N
Atlantic E-W	151	7.4	205	13.0	195	13.7	16	9.1
Atlantic N-S	28	45.5	41	18.0	116	29.7	5	34.8
East Indies	161	55.5	195	65.5	183	66.4	38	94.8
Coastal	14	4.1	11	14.8	28	24.3	2	0.0
Arctic	44	0.5	0		13	0.3	4	4.8
Other	3	57.7	16	60.7	6	24.2	0	
Total	401	28.8	468	37.0	541	34.2	65	60.6

Note: E-W and N-S denote the general orientation of Atlantic traverses, in compass points. The sampled number of logbooks per country and route precedes the observational average.

ical. Under extreme conditions, explorers like William Baffin and Thomas James set up their instruments on inhospitable ice floes in Hudson's Bay to take ten or more consecutive readings. They obviously were aware of the importance of new magnetic data from regions previously little explored.[14]

However, average vessel speed on the oceans (about 175 km per twenty-four hours of sailing), in combination with modern reconstructions of the spacing between isogonics, supports the assumption that one measurement series per day would usually suffice to note changes in declination of a degree and over. Another indication of session frequency can be extracted from navigation texts. Richard Norwood's 1637 *Seaman's Practice* stated that "it requires dayly, or once every two or three dayes halfe an houres worke."[15] A somewhat more lax stance was advocated by Georges Fournier (1676), suggesting no week should pass without determining magnetic declination. Not much had improved by the early decades of the eighteenth century; both Wilson (1723) and Harris (1730) advocated "once in two or three days if opportunity offers." To avoid confusion, the value allowed for in steering was supposed to be written down every day. A more diligent approach of daily measurement became commonplace later in the century.[16]

If the different average length of voyages to various destinations is kept in mind, the mean of the totals of all observations made during a complete voyage may serve to compare behavior on board ships of different countries, sailing to various destinations. Table 6.2 gives an idea of the number of measurements taken on several types of routes, by dividing the total number of observations by the number of sampled logbooks (in each country) that contained magnetic data. When all results from small samples are ignored, there are only a few substantial differences worth noting. In terms of routes, East India voyages showed the highest means, followed by journeys traversing the Atlantic in mainly north-south direction. Triangular slave trade (Europe–Africa–West Indies–Europe) and east-west Atlantic crossings featured a decidedly lower incidence of measurements. Ships that did not leave the continental shelf displayed very mixed results, as did the special cases (surveys and discovery voyages, Pacific traverses, and such). Uniformly low were the counts on journeys in the North Atlantic, mostly on whaling ventures (to Spitsbergen, Nova Zembla, and Jan Mayen) and Danish naval patrols. As stated above, part of the observed differences may be explained by unequal voyage duration; a one-way trip to the whaling grounds took little over one month, whereas an outwardbound East India run could easily last more than half a year.

An important issue from the standpoint of current geomagnetic modeling is the accuracy of historical measurements; even at the time itself it was known that the magnetic reading was subject to both observational and instrumental error. The former has to do with atmospheric conditions, refraction, parallax, unsteadiness of the ship, deviation by nearby iron, human error in performing the task, and other suboptimal circumstances. Instrumental error concerns the technical limitations of the compass itself, involving size and graduation of the card, intensity of needle magnetization, construction faults, friction, imbalance, and other defects.

The problem of scale can be traced back to the late sixteenth century. Prior to the introduction on board ship of the observation compass, with a card graduated in degrees, mariners had to make do with the steering compass to estimate magnetic declination. Besides having to rely on a small card divided into quarter points only, these instruments also lacked a sighting mechanism, reducing the practice to a crude eyeballing exercise.[17] Several early observations are thus recorded in points or parts thereof, instead of degrees. Some merchant vessels even continued this notation deep into the eighteenth century, even though by that time they probably employed bearing compasses to obtain the result.[18]

When and where dedicated sighting devices became common, a transfer to measurement in degrees usually followed. With the improved accuracy, differences became apparent between separate observations, made consecutively or in parallel with different compasses. The magnitude of a perceived discrepancy large enough to merit special attention in the log can be shown to shrink with time. Explorer Luke Foxe, near the Orkneys in 1631, had trouble crediting two observations four degrees at odds with each other. Navigators on board the *Royalle* in 1682 experienced the same range of values. A separate marginal note highlighted a difference of five degrees measured in 1696 by the crew of the *Vosmaar*. On the other hand, the log keeper of the VOC's *Banda*, traversing the Indian Ocean in 1636, found a difference of two degrees already noteworthy, while an anonymous extract of a French voyage from 1778 mentioned a pair of single amplitudes one degree and four minutes apart as doubtful on account of their "large" discrepancy.[19]

Of course, an untrustworthy observation could be supplanted by a new and better one as soon as opportunity arose, which sometimes required several days' patience.[20] Alternatively, the mariner could attempt to improve the adopted procedure to obtain more reliable results. Navigation texts advocated several techniques to obviate or at least reduce the incurred errors. Pierre Bouguer (1729), for instance, suggested restricting observations to stars on the hour circle of six o'clock, while a certain J. Mudbey, writing to the Board of Longitude in 1793 on matters of navigation, advised a standard procedure of taking six sights, "turning the compass round each time, to be sure the card does not hang, and to work the mean of those six sights." A rare instance of a logbook describing such a practice (alternating cast shadow with direct alignment by eye) is found in the record of the Dutch Admiralty's *Bellona* (1789).[21]

When sailing in company, crews on all vessels could share the obtained data. Comparisons of results occurred, for instance, on the *Banda* and *Zeeland* (1636), the *Sunderland* and its two sister ships (1710), the *Montagu* and *Prince William* (1745), the *Warwick* and *Intrepid* (1774), and a Dutch naval squadron comprising the *Thetis, Mercuur, Bellona,* and *Zwaluw* (1792).[22] A navigation tract by Verqualje (1661) suggested a different method, stipulating that compass error could be detected by carrying two or three instruments on board the same ship. This opened up the possibility of a reduction of error through taking either the mean (the statistical average) or the median (the middle observation of an ordered sequence) of several near-simultaneous observations. An entry in the 1682 log of the *Royalle* once more is telling: "The whole voyage we have relied upon several variation compasses, sometimes three, or five, but most often four, and . . . I've always taken the average of said compasses."[23] The Danish *Dronningen af Danmark* (1742–44) kept separate score of compasses number 1 and 2, while on the *Flore* (1771–72) the type or brand name identified the instruments. The Dutch East India Company logs of the *Amsterdam* (1632–33) and *Banda* (1636–37) simply mention three observation compasses used at the same time.[24]

Indeed, navigators on several ships made a careful analysis of performance differences, mostly between the compass officially issued by admiralty or company, and an instrument in personal possession, which often proved superior. More interesting from the perspective of cross-border spread of technology are the Levantine compasses tested against the Dutch official issue on the *Huijs ter Duijne* (1696), as well as the "English" compass versus the Dutch (on the Dutch ships *Eenhoorn,* 1698; *Goidschalkoord,* 1737; *Brederode,* 1741; *Herculis* and *Polanen,* 1750) and versus the French standard instrument (on the French vessel *Flamand,* 1783).[25]

A more statistical approach compares individual measurements' *standard deviations* from the calculated daily *mean.* The mean is simply the sum of all observations divided by their number. The standard deviation is a measure of the scatter of values around the mean. It is commonly defined as the square root of the *variance,* which is itself calculated by taking the data's squared residuals from the mean, and dividing their sum by one less than the number of observations. The main conclusion, drawn from a sample of 18,918 readings made on 8,491 days, is that English and French individual results tended, on average, to remain far closer to the central value than those obtained from Dutch and Danish sources, as table 6.3 makes clear. More detailed analyses for different epochs and countries furthermore show that the English improved their rather average performance of the first half of the seventeenth century to below 0.3 degrees in the subsequent hundred years, followed by a mild decline in accuracy thereafter. The French likewise improved their record from a standard deviation of around 0.5 (in 1650–1700) to less than 0.4 degrees by 1800. Over the same period, the Dutch struggled to push precision back from almost 0.7 to a still mediocre 0.5 degrees, while the Danes appear to have lost ground from 0.5 to over 0.6 degrees during the eighteenth century.

This leads to the intriguing question of what caused these differences in attained precision. Two factors immediately come to mind: the measurement practice and the instrumental error. As to the first possibility, the navigator's options have already been outlined: he could decide to take double or single amplitude readings, or resort to azimuth sightings if his compass visors allowed this. He could also either be content with a single obtained value per day or seize every opportunity to determine declination and take a median or

Table 6.3 *Sample Size of Multiple Measurements and Standard Deviation from the Daily Mean, Per Country*

Country	Sample Size	Standard Deviation
England	4,597	0.357
France	3,966	0.393
Dutch Republic	8,040	0.498
Denmark	2,315	0.601
Total	18,918	0.461

mean from each series of readings. The larger the sample size was in each instance, the more the individual errors would be reduced. Regrettably, the analysis of observation frequencies has just shown that very little difference existed between countries regarding the number of readings taken on a particular route. In fact, one of the few places where observations were processed more often than average was on board Danish East Indiamen, and precisely logbooks from this nation have the worst record in the attained accuracy.

The introduction of azimuth observations can similarly be ruled out. If shooting the Sun above the horizon would have significantly increased the measurement precision, one would expect standard deviations to go down in the latter half of the eighteenth century, whereas they actually increase. A second counterargument comes from French quarters; as their navigators did not rely on azimuth observations to any substantial extent, they should have been heavily penalized, which does not at all match their high ranking, being second only to the English.

This leaves the compasses themselves as the most likely cause. Apparently, both English and French instruments offered the navigator better results across the board. The advantage was already established by the early seventeenth century, and was retained during the following two hundred years, despite several innovation attempts undertaken by the Dutch and the Danes. Perhaps the drive and rewards for compass improvements were greater in England and France, and officials and mariners accepted them more readily than in Denmark and the Dutch Republic. Alternatively, the level of craftsmanship may have been slightly ahead in the first two countries. One possible explanation concerning the difference between Dutch and English instruments is the level of standardization in production. In England, the Royal Navy initially contracted a single supplier, and appointed its own compass maker after 1728. By contrast, both the Amsterdam Admiralty and the Dutch East India Company employed several firms, and no monopoly was ever granted. Although the resulting competition may have helped to keep prices low, it would also have led to more variability in quality. Sadly, not enough is known of the practice in France, Denmark, the English East India Company, and private maritime enterprises in general, to account for the clear differences in standard deviations around the daily mean there. However, the instrument itself was not the only factor in the equation. External influences could also seriously affect the outcome of measurements.

Impediments

Around the mid–sixteenth century, Neapolitan natural philosopher Giambattista della Porta engaged in some rather peculiar experiments, "breathing and belching upon the loadstone after eating of garlick." To his surprise, this lack of table manners did not have any perceptible effect on needle magnetization. Even "when it was all anoynted over with the juice of garlick, it did perform its office as well as if it had never been touched with it," he wrote in 1558.[26] He was investigating a claim by classical authors that garlic would annul any nearby magnetic action; some attributed the same effect to onions too. Practical navigators, however, burdened by more mundane concerns than scholasticism, had little time for these imagined threats to compass reliability, as Porta himself confirmed when he quizzed them on the subject: "When I enquired of mariners whether it were so that they were forbid to eat onyons and garlick for that reason, they said they were old wives fables and things ridiculous, and that sea-men would sooner lose their lives then abstain from eating onyons and garlick."[27] William Barlowe, a few decades later, similarly made short shrift of these notions, "rust being the greatest enemie that the touche of the stone can haue . . . being farre more noisome then garlike or oile, and all the rest of those fondly surmised conceites."[28]

A real and far more direct threat to the instrument's operation was the weather, in several manifestations. The problems of clouds and haze obscuring the skies formed a never-ending source of logbook complaints. In particular areas such as Hudson's Bay, large temperature differences between water and land caused fog banks to form frequently year round. Humidity also induced rust, which increased needle friction.[29] Moreover, wind and waves sometimes conspired to keep compasses in such a stir as to make observation impossible, "the motion being too great for them to stand to any nicety."[30] This was particularly troublesome on small vessels. Gimbals are better able to counteract slow and even ship movement than sudden displacement by gusts of wind and waves crashing against the hull.

A far more spectacular type of weather interference was lightning. Officers of a small Dutch fleet exploring Australian waters in 1705 afterwards reported: "On the passage from Timor, the compasses were on the sixth of March affected by the thunder and lightning to such a degree that the north-end of the needle pointed due south, and was brought home in that position."[31] The intense electromagnetic disturbance of a nearby strike is certainly powerful enough to induce new, and alter existing magnetization, an effect which raised the curiosity of the Royal Society as well. A paper submitted there in 1684 spoke of a similar incident off New England: "The north poles of several compasses were changed south, and always continued so. The north pole of one compass was turned west, but lost its virtue in some time after."[32] The 1749 tale of the ship *Dover* is one of the most extensive, as was the havoc wreaked on board, the bolt temporarily paralyzing and blinding part of the crew, destroying main mast, mizen, and most sails, causing structural damage below decks (including a leak), and reversing needle orientation of the four compasses carried. A little later, they lost their magnetization altogether.[33]

In shallow seas, near volcanic islands, and in some ports, crustal magnetization could affect needle readings in a milder manner. Beneath the bottom of Hudson's Bay lie large bodies of iron ore; Dutch sea-atlases marked a reef on Finland's south coast as disturbing the compass within a mile's distance, and captain Foxe held strong suspicions regarding the south tip of Greenland: "Cape Farewell, I holde for certaine, doth attract the Magnet more suddainly . . . then any knowne cape in the world, as did appeare in all this voyadge."[34] But for most of an ocean crossing, the seafloor is several kilometers below the surface, and crustal anomalies tend to be of limited magnetic force, and thus can generally account for a deflection of about one degree only. Sudden jerky needle movements and continual reorienting for hours on end was incidentally observed far away from the solid crust, and thus had to be of a different origin. The most likely candidate is a magnetic storm, enveloping the Earth in a shower of charged particles from the Sun. The VOC's *Heemskerk* in 1642 experienced needle play ranging over eight points (90°) south of Australia, the crew blaming some undiscovered mine of lodestone nearby. A different explanation was put forward after a similar incident aboard the *Vrouwe Elisabeth* in 1766; the author of the logbook then incorrectly assumed that the ship was at that moment directly over the south magnetic pole. The textbook by Millet Dechales (1677) assured mariners that this unpredictable phenomenon never lasted long, but did not supply any remedy. George Robertson, master on the *Dolphin* (1768) sailing far into the South Pacific, and failing to find any plausible source on deck for his needles' constant straying, decided to "keep a very strick look out all this day in hopes of seing some land, or birds which is a sure sign of land being near but saw neather."[35] This was after opening and cleaning the instruments, comparing the steering with the azimuth compasses, and checking the binnacle for the presence of metal objects.

Needle deflection due to the retained magnetic field of nearby iron is called magnetic *deviation.* This is due to the magnetic properties of "hard" and "soft" iron. The former is not easily magnetized, but will retain its magnetism permanently, while the latter more readily realigns with any transitory magnetic field. All materials that contain iron possess both hard and soft qualities to some degree; their ratio determines their actual magnetization properties.[36] As a result, hard iron in the vicinity of a compass exerts a constant deviating pull, while the effect of other magnetized objects will be dependent on the ship's orientation relative to the local geomagnetic field.

Compass bowls made of brass mixed with iron particles could certainly explain strange needle behavior on a number of occasions. Careless construction and repair of a (wooden) compass box or binnacle with iron nails caused similar problems.[37] The two steering compasses inside the binnacle could be placed too closely together, so their needles would experience deviation by their neighbor's magnetic field. Other iron near the binnacle could likewise contribute— for instance, iron pump sticks, rudder supports, and gratings.[38] Metal parts of clothing (buttons, buckles, clasps) and other portable items (knives, keys, tools) also occasionally lured the needle away from magnetic north.[39]

Compass observations had the advantage that the instrument could be set up anywhere on deck, using a tripod, a stool, or a rotatory platform (of course, solely made of nonmagnetic materials). But because placement would vary, navigators had to remain conscious at all times of potential agents of deviation in proximity. Among the most frequently cited are pieces of artillery. The very first such report of inconsistent magnetic readings dates back to 1538, when explorer João de Castro identified a ship's gun as the disturbing element. This cautionary tale became part of his *Roteiro de Lisboa a Goa,* and has doubtless alerted many of his colleagues to the danger. Over a century and a half later

(1696), Edward Harrison reiterated the maxim: "Suffer no great guns or other iron too near your compasses."[40]

At other times, differences between values measured at various locations on deck remained inexplicable, as was found on board the *Chandos* in 1690: "Note th.t the observ. was taken on the larbord side, and I have taken notice of a considerable diff. in changing the compass from side to side."[41] The crew of the *Philibert* (1734) similarly discovered discrepancies between observations made on the deck of the poop and the forecastle. Hydrographer d'Après de Mannevillette added a note in his 1765 sailing directions to the East Indies, ascribing the error there to the ship's anchors, and proposed the poop as the most convenient location.[42] A solution put forward by surveyor John Churchman in 1794 is known as *swinging the ship*. This procedure tracked the influence of transient magnetization by checking and recording needle deviation on all headings while turning the ship full circle around a vertical axis. This would become standard practice in the nineteenth century, when iron and steel came to make up a substantial part of hull and skeleton, and deviation became the most substantial problem associated with the magnetic compass.[43]

The last impediment to proper observation of magnetic declination to receive attention here has to do with the fact that, in high latitudes, the local horizontal geomagnetic force is small relative to the vertical part. Because most of the magnitude of the three-dimensional vector is there seated in the vertical, the needle dips downward, which increases pivotal friction. At the same time, the aligning force in the horizontal plane is less than at lower latitudes. The compass is said to become "numb," the magnetized arrow being not very sensitive to directional change, and slower in regaining its orientation after displacement.

Antonio Pigafetta's narration of Magellan's circumnavigation (1525) speaks of an episode in the South Pacific, when the captain-general alerted all navigators that the needle did not receive as much force as in its own quarters (meaning the northern hemisphere). *The Worldes Hydrographical Discription* (1595) by navigator John Davis of Sandridge contains a passage on the region between the parallels of sixty and eighty degrees north latitude, where "quicke and uncertayne variation of the compasse" was to be expected.[44] Captain Foxe, during his exploration of Hudson Strait in 1631, furthermore wondered whether "the sharpnesse of the ayre, interposed betwixt the needle and his attractive point, may dull the power of his determination; or here may be some moun-

taines . . . whose minerals may detaine the nimblenesse of the needles moov-
ing to his respective poynt."[45] One could compensate the needle's downward
tilt with a counterweight, ranging from a bit of wax attached to the underside
of the card to sophisticated mechanisms that allowed variable adjustments.
Such measures at least prevented the card from scraping against the glass lid,
but the lack of horizontal directive force remained. In combination with high
refraction, frequent bad weather, poor charts, and all dangers associated with ice,
the desensitized compass constituted one more reason why high-latitude ventures tended to present navigators with greater challenges than a Caribbean run.

Fixing the Needle

Before the introduction of the steering wheel in the early eighteenth century, the sternpost rudder was moved by a whipstaff, basically a lever pivoted
near its midpoint. Its lower end was attached to the tiller, the upper part held
by the helmsman; leverage could be improved with the aid of a tackle. In front
of the helmsman stood the binnacle, usually containing two compasses. In addition to the one between the mast and the great cabin, a second binnacle was
situated on the topmost deck of the poop. The helm was never left unattended
while the ship was out of port. Compass and wind direction helped to maintain a set course, which was ideally recorded every half hour.[46]

The steering card being divided only in points and parts thereof, the ship's
headings were kept using the same division. The maximum attainable accuracy was dependent on ship type and weather. Under favorable conditions, a
quarter point ($2°49'$) was within the capability of an able helmsman, while
some authors of navigation manuals deemed half a point ($5°38'$) sufficient or
the highest attainable. Potential error affecting a magnetical course could, unfortunately, come from many corners. The bedeviling specter of deviation has
just been reviewed. Misdirection by a couple of degrees or more could also result from uneven magnetization of oval and lozenge-shaped needles, or the
binnacle being oriented at an angle with the fore-and-aft line of the ship. A
more pervasive and labor-intensive problem was magnetic declination. As
stated earlier, one has to distinguish between the value measured (in degrees
and minutes), and the correction applied to (1) the steered course (often in
quarter points), and (2) the ship's heading in dead-reckoning calculations (in
degrees).[47] Like his colleagues, the helmsman on ocean-going vessels appears
to have been very much aware of local declination, evident in his personal ad-

justments en route. These could take several different forms. The one most often applied occurred only inside his brain, and has therefore left no trace in the records. Mental correction could immediately accommodate whatever field features were encountered; its main drawback was its susceptibility to human error. Perhaps this is why two physical solutions have evolved, placing the correction within the steering compass itself.

From the second half of the fifteenth until well into the nineteenth century, West European coastal shipping employed a mechanical solution that dispensed with the need to either observe or correct for changing needle orientation. It consisted of permanently fixing the needle at an angle to the north on the card. This offset was meant to counteract local magnetic declination as measured in the home port, or as averaged over the region crossed to reach regular destinations. Permanently slewing the needle relative to the fleur-de-lis of course only worked over a small area, and during a short span of years. Moreover, it was precisely these two limitations that caused much confusion. Regarding the geographical aspect, in every region compass makers compensated for the local declination, and navigators and hydrographers drew charts on that basis. In 1581, Robert Norman listed no fewer than five types of "corrected" compasses and charts, each suitable only in specific waters (see table 6.4). Several authors subsequently copied this list with some alterations during the seventeenth century.[48]

The second cause for concern lay in the change in magnetic declination with time. Although not yet acknowledged at the turn of the seventeenth century, it was all the same reducing northeasting at variable speed all over the continent, foretelling the coming of northwesting to West European shores in the second half of the seventeenth century. Yet instrument makers seem to have followed tradition rather than magnetic forecasts, at least according to numerous nautical texts. Thomas Harriott in 1595 spoke of "the common compasse, whose wires stand half a poynt to the eastward," as did William Borough the year after, Richard Polter in 1605, and William Baffin in 1613.[49] In a letter on the subject to Samuel Pepys from 1697, Astronomer Royal John Flamsteed wrote: "Some friends of mine assure me they have seen old sea compasses with their needles so placed, and I have by me a collection of some variations . . . taken by such compasses."[50] On the other hand, longitude-finder Zachariah Williams recalled in 1745 that a number of these devices had their needles fixed to negate north*westing*, indicating that compass makers may have eventually caught on to the fact that the field was changing. Master Martin Pring's logbook of

Table 6.4 Robert Norman's Five Types of Fixed-Needle Compasses

Compass Origin	Compensation for NE	Suitable Area
Italy	0	Levant
Flanders, Danzig	¾ points (8°26′)	NW Europe
Flanders, Danzig	1 point (11°15′)	NW Europe
England	1½ points (16°53′)	Russia
Spain, Portugal, France, England	½ point (5°38′)	W Europe

Nicholas Downton's 1614 voyage to the East Indies furthermore holds an interesting remark regarding oceanic practice: "From England to the Canaries we used our Chanell compasses, and from thence our meridionall. Whiles we used our Chanell compasses we gave not any allowance for the variation, which afterwards we did in all our courses." This remark once more underlines the fact that navigation on and off the continental shelf constituted radically different worlds.[51]

In France, a single shift of magnetic opinion appears to have occurred around the 1620s or 1630s. Jean Rotz, writing in 1542, grouped Scottish, English, Flemish, and French compasses together in having the wire fastened to the fly at half a point east of north. Similarly, two Newfoundland charts published in 1612 and 1613 by explorer Champlain assumed just over five and a half degrees northeasting. In 1644, however, Petrus Herigonus reported that French common compasses used in coastal navigation then more or less followed the Flemish offset of eight to nine degrees. Note that the easterly correction had increased, whereas the actual observed values at the time continued to drop. Scattered French references to the fixed needle made further appearances up to the early eighteenth century.[52]

A possible change in Dutch practice can be tentatively dated to the 1620s. From the 1580s onward, the Northern Netherlands had followed their southern counterparts, personified in Michiel Coignet, who had taken the value then measured at Antwerp (9° NE) as the norm. Dutch ships bound for the East in 1598 carried both meridional and so-called "Hollands" or "Amsterdam" compasses, the latter having the needle fixed to counter two thirds of a point northeasting (7½ degrees). The 1623 log of the *Wapen van Delft* instead described a comparison of two other types of fixed-needle compasses: from Amsterdam (at 7°06′ NE) and Rotterdam (at 5°30′ NE, or about half a point). Some Dutch navigation manuals, published in the ensuing decades, came to describe common compasses as having "over half a point," while others suggested "almost one point," and "two thirds of a point." All of the above references cast doubt on whether Dutch compass makers adhered to a single standard after the 1620s. The practice would continue for over a century in coastal traffic; a 1762 logbook from a merchant vessel sailing to the Baltic still steered with a "contemporary compass" without allowing for any magnetic declination. At the time, reigning northwesting locally exceeded twenty degrees.[53]

The above-mentioned textbooks did not necessarily endorse the practice of fixing the needle, but merely reported its existence. Some authors actively opposed the concept, for various reasons. Bourne stressed spatial variability of declination, Norman and Sarmiento the possible discrepancies between charts and compasses. Norwood's *Seaman's Practice* (1637) is remarkable in that it pressed into service the argument of the field's change with time only two years after its discovery. Colleagues Millet Dechales (1677), de Groot and Vooght (1684), and Newhouse (1701) later voiced similar opinions.[54]

The Rectifier

This instrument consisteth of two parts, which are two circles either laid one upon, or let into the other, and so fast'ned together in their centers that they represent two compasses, one fixed, the other moveable; each of them divided into the 32 points of the compass, and 360 deg. and numbred both ways, both from the north and the south, ending at the east and west in 90 deg. The fixed compass represents the horizon, in which the north and all the other points of the compass are fixed and immovable. The movable one represents the mariners compass, in which the north and all other points are liable to variation.[55]

Thus wrote James Atkinson, teacher of mathematics, in his 1707 *Epitome of the Art of Navigation*. The device he was referring to could be made out of paper, pasteboard, wood, brass, or any other suitable material.

The *rectifier* constituted the second method of physical correction in steering. It was slightly more sophisticated than the fixed-needle solution in that it could accommodate, and indeed required, variable adjustment, based on current information on local magnetic circumstances. This could come from direct observation, personal experience from previous voyages, logbooks of predecessors along the same route, or official sailing directions stipulating a certain declination to be allowed for. But whatever the source, the method was the same, and merely involved setting the inner fly at the specified angle to the outer. The two cards then formed a circular translation table between magnetic and true courses, which functioned as a reference at the helm; it immediately converted any requested true heading (outer rim) to the corresponding magnetic heading (inner rim). Needless to say, the steering compass had to have its needle fixed north-south, else new error would be incurred.

Evidence of its use on the oceans in this naked form is sketchy. The English nautical textbooks of Sturmy (1669) and Speidell (1698) contained a description of it. A brass "compass straightener" (*compasrechter*) plus instruction manual also figured twice in the 1655 list of navigation instruments signed for by Dutch masters upon departure for the East Indies. The next edition twenty years later, however, no longer made mention of it, witnessing its silent demise.[56] Explicit references in logbooks are sadly lacking altogether. For a possible explanation, one may have to take a closer look at steering compasses themselves. Given the idea of the rectifier, it was only a small conceptual step to placing this double fly not next to the compass, but directly on top of the needle. In that case, the outer rim no longer needed to be visible during normal operation. Instead, the rectifier became a system of two cards of equal diameter, the lower borne by a fixed needle, the upper able to be turned at will to counteract magnetic declination. Allowance was thus made purely by construction, the helmsman no longer having to convert magnetic to true, either in his head or by means of the separate discs described by Atkinson above.

This alternative solution came to be known in England as "mouable fly," in France as "double card" (*rose double*), and in the Dutch provinces as "shifting card" (*schuivende roos*). English sources attesting to its use are limited in number and time span. Thomas Harriot's mathematical papers discussed the consequences of using both fixed-needle and rectified card in 1595. Hudson, on his 1610 attempt to find the Northwest Passage, furthermore described the crossing of the Atlantic agonic near the Orkneys, where he "set the north end of the needle, and the north of the flie all one."[57] But the practice was not to last; twenty-one years later, captain Foxe on a similar adventure in the north "sailed all by meridian compasse," the needle's variation being accounted for in the dead-reckoning.[58]

French endeavors appear to be confined to the second half of the seventeenth century. Denys (1666) actually recommended use of the double card, whereas Millet Dechales eleven years later did not particularly care for it. Berthelot in 1701 was among the last French writers on navigation to describe this physical correction; by that time, adjustment through calculation had already gained much headway. Of the few extant French logbooks prior to 1700, four (all from the Compagnie des Indes) hold explicit remarks like this one: "We have turned our compass card for 15 degrees of variation northwest, estimating that there has to be 15 to 16 degrees of variation at this location."[59]

Danish card rectifying seems to have been limited to the 1720s and 1730s. This assumption is based upon occurrences in two logs of the Danish Guinea Company, two of the Asia Company, and four of the Danish Navy. These were, however, even then the exception rather than the rule.[60] Contrastingly, Dutch traces of the double fly are ubiquitous and reach far into the nineteenth century. The "shift compass" (*schuifcompas*) figures in logs from the earliest East Asian ventures and was standard equipment throughout the East India Company's history, surviving all revisions of required navigational instruments. Early evidence of allowance in Dutch source material from the merchant navy appeared in 1707, from the slave trade in 1721, from the admiralties in 1737, and from whaling in 1784.[61]

Dutch navigation textbooks often even had separate sections or chapters devoted to correction by fixed-needle and rectifier. The discussion by Maartensz (1701) provides the additional detail that the top card could have a small indent to facilitate turning grip. In a passage somewhat similar to that of Pring's logbook quoted earlier, sources from the Dutch East Indiaman *Noortbeek* illustrate common practice for ocean-going vessels: upon leaving the last port before a crossing, dead-reckoning was formally started anew "with a compass set to [X] degrees northwesting." During the voyage, each change was usually faithfully recorded, using phrases like "the compass card(s) reset from [X] to [Y] degrees," allowing permanent monitoring of local allowance made. It is by virtue of these constant updates that such compasses were frequently called "straight-pointing" (*regtwysende compassen*), as opposed to "misguiding" ones (*miswysende compassen*) used for observation purposes. Some logs (such as from the Dutch Admiralty's *Pollux*) even contained a preface certifying that all observations, winds, and courses were recorded using the corrected instruments.[62]

Of more general concern regarding compass allowance is the interaction between observed values and those corrected for. In some instances, the relationship was very direct: the measured deflection was compensated exactly to the minute.[63] Dutch practice was once again out of step in often adhering to a system of shifts of fixed size. In the early seventeenth century, these tended to

CHAPTER 7

Plotting the Third Coordinate

During the seventeenth and eighteenth centuries, geomagnetic field data were primarily of interest to three groups of professionals: natural philosophers, hydrographers, and navigators. The philosophers dealt mostly with geophysical theory and its application to determining longitude; the navigators were concerned predominantly with its practical implications at sea. Hydrography bridged the gap between the two. Its land-based practitioners relied on the maritime community for information to make and revise charts and sailing directions. Several also taught and examined masters in navigational matters. Some, like Cornelis Lastman and d'Après de Mannevillette, had personal navigating experience from a previous career, but this was no prerequisite. The Blaeu dynasty of cartographers, for instance, did not engage in extensive seafaring at all. Similarly, William Barlowe's treatises, published around the turn of the seventeenth century, are among the most expert available at the time, yet in the preface to his 1597 *Navigator's Supply* the author frankly admitted: "Touching experience in these matters, of my selfe I have none. For . . . by naturall constitution of body, even when I was yong and strongest, I altogether abhorred the sea."[1]

Hydrographers and examiners of navigational knowledge worked for East India companies and for navies, to improve navigation in the widest sense. This could entail establishing faster or safer routes to various destinations; determining the proper coordinates of islands, landmarks, and ports; cataloging of, and alerting mariners to, potential dangers; charting coastlands and depths; analyzing the constant and seasonal patterns of winds and currents; setting standards for proper instruments and measurements made therewith, and last but not least, tracking the constantly changing geomagnetic features all over the globe. In order to acquire the necessary body of knowledge to execute these tasks, detailed instructions were given to mariners, thus ensuring a steady stream of high-quality data. In 1612, Captain Thomas Button received the following advice upon embarking on his quest to find the Northwest Passage to China: "As often as occasion offers it selfe . . . let some skilfull man, with good instrument, obserue the eleuation, the declination, the variation of the compasse."[2] Prior to a similar venture over a century later (1741), the Lords Commissioners of the Admiralty ordered Captain Christopher Middleton: "You are there to make the best observations you can of the height, direction and course of the tides, bearing of the lands, depths and soundings of the sea and shoals, with the variation of the needle."[3]

The natural carrier for such intelligence was the logbook. East India companies and navies therefore often required their officers to surrender either the log itself or a detailed extract upon arrival at the destination, or after returning to the home port. Sanctions for neglecting this duty could vary from one to several months' wages being withheld. Payment could even be entirely conditional on the handing in of all papers concerned, as was the case for officers in the Dutch Admiralties from 1749. In the second half of the eighteenth century, the English Navy even had a separate form designed to facilitate and standardize the process. It contained columns to hold specific information on place and time, latitude, longitude, and variation, the best directions for sailing into or out of ports, marks for anchoring, wooding and watering, provision and refreshments, fortifications and landing places, trade, shipping, and so forth. The large sheets were to be securely sealed "to prevent their being exposed to the inspection of any person through whose hands they may pass," and sent to the secretary of the Admiralty Board at the end of every six months.[4]

The strategic value of maritime knowledge as evinced above was already apparent in the 1633 instruction of VOC hydrographer Willem Jansz Blaeu, who was responsible for all logbooks and journals being submitted to the East India House in Amsterdam. Nevertheless, some information did end up in foreign hands, either in the ready form of charts, or as rough voyage data. For his *Mémoire sur la Navigation de France aux Indes,* d'Après de Mannevillette ploughed through over 250 logbooks, compiling the relevant notes of each in a separate extract. He also made use of non-native sources, such as the logs of the *Grantham* and the *Egmont.*[5] Magnetic declination was frequently of particular interest, as testified by de Mannevillette's efforts to gather thousands of such observations in special notebooks. Similarly, in a letter to the Ministry of the Navy from 1780, French scholar Le Monnier requested to be given access to the log of Lozier Bouvet, who had traversed Australian waters in the late 1730s, because "there is some interesting material on the variations of the magnet . . . in the log."[6] Five years later the same ministry received a letter from explorer d'Entrecasteaux, writing from the Cape, which stated: "The great advantages of knowledge of the variation of the compass . . . made me decide to inform you . . . that I have observed [it] in False Bay, so it can be conveyed to vessels bound for the Cape: this variation changes every year; it is necessary that measurements are communicated yearly."[7] Similar English sentiments are in evidence in John Malham's *Naval Gazetteer* (1795), a massive, two-tome, alphabetized listing of all known coasts, comprising over a thousand pages.[8]

Sailing Directions

Hydrography was very much a two-way street. The process bears a marked resemblance to geomagnetic modeling: bulk amounts of raw maritime data are compiled, sifted, weighted, and combined into a single global estimate of local conditions. In the historical case, the output took the shape of charts and sailing directions that were made available to the sea-going professions. The information contained therein mostly came from navigators at sea, who charted new coastlines and corrected existing descriptions in their logs. Hydrography and navigation thus formed a self-improving symbiosis.

It was in classical times that mariners first sought to lay down in writing the succession of courses and distances necessary to reach distant harbors, previously passed on only by oral tradition. The *pilot book* was, as the name implies, primarily a coastal guide, containing descriptions of headlands, anchorages, port entrances, and sometimes currents and tides. The given headings were initially magnetic courses, that is, uncorrected for local declination. This practice started to change once Iberian explorations inaugurated the era of oceanic navigation. Growing awareness of needle deflection from geographic north at sea has been touched upon in the previous chapter, as were the roteiros, which similarly became widespread during the sixteenth century. These directions for travel off the continental shelves differed from coastal route guides in a conscious attempt to store and pass on notions of local field characteristics. Consequently, headings were true, and the noted values of declination served to reflect a lifetime of the author's experience. However, their idiosyncrasy posed a problem; an individual's collected observations over decades, of a changing phenomenon still deemed time-invariant, led to serious inconsistencies, both between roteiros and with the outside world. Initially, the problem was solved by combining many different authors in a single compilation. Most available information on a region was then at least at hand, so navigators could consult a host of predecessors, even if these did not always sing in unison. The five-volume *Itinerario* is a good example, edited by Jan Huyghen van Lin-schoten, based on the work of Affonso, Rodrigues, and others (1570s–1580s), and itself going through five Dutch editions (1595–1644), as well as one English (1598) and two French ones (1619, 1638).[9]

In 1625, Rodrigues's magnetic measurements were also published separately in a navigation textbook by Manuel de Figueiredo, and as late as 1703 a French logbook referred to a roteiro by da Motta. By that time, the Dutch had long su-perseded the Portuguese as the prime source for such knowledge. Dutch coastal directions appeared in French texts by Glos de Honfleur (1675) and Fournier (1676), and after 1655 the Dutch guidelines for sailing to the East In-dies became available in printed form for wider distribution, despite official se-crecy. Some copies are now found in the archives of the French Navy, while others became a part of the English East India Company's paper legacy.[10]

In its early years, the latter organization tried to impose a similar scheme of standardized sailing directions. Moreover, from 1614 all logs were copied, and charts were compared with those made by the Portuguese. But already four-teen years later, its Court of Committees abandoned the procedure, giving the master only his commissioned destination instead. Left to their own devices, English navigators either kept private records for reference (seventeenth cen-tury), or relied on Dutch and French material (eighteenth century). John Seller's maritime atlas, based on older Dutch material, has been mentioned earlier. Later, the work of the Dutch cartographers' dynasty of van Keulen sim-ilarly obtained a large readership across the Channel. Other examples include d'Après de Mannevillette's French *Mémoire,* appearing in English translation in 1769, while the 1778 *Seaman's Guide* by John Diston was based in part on "the latest and best surveys, of the English, French, Dutch, and Danes. The courses by the compass, and distances from place to place with the variation lay'd down as observed at this time."[11]

By then, the Dutch East India Company had all but gained complete nu-merical dominance in sailing directions to, from, and within the East Indies, with major printed publications in 1746, 1748, 1768, and 1783, and minor reprints with updated figures for local magnetic declination at several in-stances in between. In addition, each individual ship also received its own "sail-ing order" (*zeilaas-ordre*), which specified its particular route in more detail. General regulations determined that only the ship's council of officers could take the decision to abandon these prescribed courses, and even then solely in exceptional circumstances. Before a formal revision could replace its predecessor, some ships served as guinea pigs, their navigators being encumbered with the requirement to submit a written report of perceived merits and flaws of the alternative route after completing the voyage. The deliberations preceding a new proposal involved the examiners of masters of all six chambers of the company, and could take years. Records of the negotiations leading up to the 1768 edition have been preserved, and underline the importance of magnetic data in the discussions; in seven out of nine articles reviewed did the bone of contention involve the geomagnetic field and its secular variation.[12]

The Positional Indicator

On 27 December 1695, the sickly crew of the English East Indiaman *America* anxiously awaited the sight of land. The log reads: "Our westing made from Zeloan whear wee see the water discullered is 231 myles . . . Yesterday a great long log of timber swim by the ship that was full of barnickles, suposed to be bloun from the Malaabar isleands or the main. Wee looke uppon as a sine not to be farr from land, which pray God send us a happey sight of, for wee are now in a uerey weak condition, not hauen tenn well men in all our company w. is this day remaining aliue, butt many of them are uerey weake, not expected to liue."[13] Floating debris and the altering of the color of the seawater were just two of the means to obtaining positional information outside the realm of dead-reckoning and celestial observation. Near the Cape of Good Hope, a certain type of seaweed would herald imminent shore leave in Cape Town, and on the Pacific expedition of Jean de Surville and Guillaume Labé in 1769–70, the scent of meadow had indicated the proximity of land. It was also generally known that specific species of birds fed within a given range from the mainland and islands. In 1704 master Twist of the Royal Navy's *Kingfisher* identified one as a "St. Helena pigeon." At the time, the ship was in fact quite near Trinidade and Martin Vaz, but on a second occasion almost exactly a year later, two more specimens did announce the welcome harbor of the eponymous island two days in advance.[14]

Some animals even surpass human navigators in accurately negotiating whole oceans. Loggerhead sea turtles hatched in the Americas engage in transoceanic migration following the nutrient-rich North Atlantic gyre to remote islands such as Ascension. Eventually they manage to return to their natal beach with uncanny precision, as recent satellite tracking has verified. These ancient reptiles are able to perform such feats using the Earth's magnetic field, in this particular case by detecting both geomagnetic inclination and intensity. Apparently, each newborn turtle learns to associate specific combinations of some field components with position or bearing.[15]

Although not consciously emulating nature, humankind did manage to harness the path-finding potential of the field in a less sophisticated fashion. Steering and dead-reckoning have already passed review. Presently, a more direct link between magnetic declination and identified locations merits attention. It hinges on the fact that navigators, registering magnetic declination time and again at a given point (either near a landmark or in a sea-lane) came to associate these values with that place. Few maritime historians have so far acknowledged the importance of this positional clue; it helped to determine the nearness of land, to approximate and to correct longitude, and, over areas of assumed regular progression of northwesting or northeasting, it sometimes even yielded a positional fix to the mile. These applications were all based on extensive seafaring experience, and have to be clearly distinguished from global longitude schemes as covered in the first part of this book. What follows is thus primarily rooted in navigational practice.

Recognition of the potential benefits of the wandering needle are easy to find in navigational texts. Consider these two examples: George Waymouth in 1604 attested that "there is no one thing in navigation seruing to more excellent use then the variation is, for by God's grace I dare undertake to conduct a shipp to any of those partes where I haue myselfe obserued the variation and the height of the pole and by them knowe inst.ly when I shall see any of those partes againe [and] that it pleased God to make it the onlye meane whereby my owne life, and the liues of all those that were with me were wholely preserued." Henry Coley, writing in the 1670s, certified in more general manner that "although it pretend uncertainty, yet it proveth to be one of the greatest helps the seamen have."[16]

The association of declination values with nearby land could be either direct or indirect. Numerous sailing directions and personal notes explicitly linked local needle behavior to way stations such as the Azores, Ascension, St. Helena, the Cape of Good Hope, Madagascar, Mauritius, Diego Rodrigues, St. Paul and Amsterdam Island, as well as various final destinations in the East Indies. Less often the sources mention places along the American and African east and west coasts. Out of sight of land, the so-called Cart Track (the Atlantic corridor to cross the equator southward) formed a steady part of most Dutch transoceanic guidelines, while various French and English sources contain ref-

erences of specific needle allowance associated with first soundings of continental shelves. Buccaneer and explorer William Dampier even preferred magnetic declination as a telltale sign of passing the longitude of the Cape above
soundings of the sea bottom there, at fifty to sixty leagues from shore. Similarly, a private sailing direction in the logbook of the English East India Company's *Susanna* (1683) remarked: "att Cape de bones Esperens we have uppon
soundins we haue [*sic*] 10d varea.," while navigators on the ship *Walpole* in
1759, skimming the Australian coast, recorded "just before we had soundings
6°33' and in sight of land 5°50['] so that keeping above 7°00m you will be clear
of all danger."[17]

In the case of small islands, the measured magnetic meridian could hold the
answer to whether the ship was to the west or to the east of it. Geographer
Nathaniel Carpenter went a step further in 1635, when he reasoned that few
places would have both latitude and declination in common; two years later,
Richard Norwood concurred, in saying that "some neare conjecture of his longitude by the variation" could aid the mariner in his voyage across the vast
ocean.[18] An unknown author edited by hand a sailing direction from the 1680s
appended to the log of the Compagnie des Indes's *Royalle*, to change the ear-
lier-printed location of the intersection of the Atlantic agonic line with the
Cape's latitude from 60 to 270 leagues west of the African main, "which is a
good marker for knowing how far one is distant from land," while Captain
George Cuming, on his way to Ceylon in 1752, stated upon measuring forty-
five minutes northeasting at 7°22' north latitude: "It is somewhat extraordinary
to find East variation for 7 degrees together in this lat.d, especially as there is
near 2 degrees West Variation off of Achein Head." Similarly, Struick's 1768 navigation manual gave the example of latitude and declination yielding a longitude accurate to within two degrees.[19]

Isogonics thus functioned as a grid reference, even though these generally
did not form straight lines. The notion of nonlinear coordinate systems is
nowadays quite common in mathematics, but it is somewhat surprising to find
such a practically minded group of professionals as early-modern navigators
able to appreciate its benefits. In the absence of reliable longitudes, latitude and
magnetic declination could alternatively pair up to uniquely define a position.
For example, a description of St. Augustine's Bay (west Madagascar), found in
the log of the Royal Navy's *Hawke* (1669), concludes with: "lat.d by good ob.
23.31 south, variation ob.d 23.32 west."[20] A chart of Réunion in a French logbook from 1720 designates the island to lie in 20°15' south latitude and 21 de-

grees northwesting.[21] When longitude did occur in descriptions, a triplet of coordinates (latitude, longitude, and magnetic declination) was sometimes used. A 1671 draft of Buss Island had all three neatly lined up, and a 1788 report on Mauritius's main harbor lists a number of observations by accurate observers: "Latitude observed in the town of Port Louis 20.9′33″; Longitude observed by means of 70 distances of the Sun and Moon on ten different days from Greenwich 57.29′15″ E; Variation of many observ.ns 17.10′ W on shore, 16.20[′] in the road."[22]

At sea, magnetic data could serve to check and even correct the longitudinal estimate. A French naval document from the mid–eighteenth century discussing a route past Rodrigues Island asserted that "the quantity of variation will let us know with ease . . . the great error that we could have in estimating longitude."[23] Two others from the eighteenth century declared respectively that knowledge of declination was of equal value as and of higher worth than dead-reckoned longitude.[24] A text on the usefulness of the compass, kept among the papers of the Paris Naval Watchmaker's Office (Horlogerie de la Marine), put it bluntly: "There are places where the quantity of the declination of the magnetized needle alone can determine the ship's longitude."[25] Several logbooks even witnessed a longitudinal correction of several degrees (that is, hundreds of miles), solely on account of encountered magnetic declination. Among them are the English East India Company's *Jewel* (1637), the Danish Navy's *Grev Laurvig* (1725–27), the French *Paquebot no. 4* (1788), and the Dutch East India Company's *Unie* (1790).[26] Such drastic revisions of estimated position normally occurred only after sighting land.

As early as 1557, Cortés had suggested making compilations of northwesting and northeasting as observed from port to port, for future reference. Awareness of an expiry date of such gathered data had yet to be established.[27] The discovery of secular variation, however, did not spell the end of the technique, but rather pressed home the need for more frequent readings and for regularly updating the figures. Textbooks by van Nierop (1676), Berthelot (1701), Dassié (1720), Steenstra (1770), and C. Pietersz (1779) all stressed both the necessity of constant vigilance and the potential rewards.[28] Particularly in the Dutch Republic, the number of references to the Earth's magnetic field in sailing directions increased dramatically in the first half of the eighteenth century. Direct experience could only serve for a couple of years, and even then only with the proviso that the chosen trajectory was the same. An example from the English vessel *Anson* in Philippine waters (1751) states: "I allow no variation in these seas, I by several observations outward bound found none here."[29]

Several authors condemned the oral communication of such information, among them Harrison in 1696 and Meynier in 1732. An equally questioned practice was for navigators to take their magnetic cues from old logbooks. Bouguer (1753) considered such negligence extremely dangerous.[30] In addition to the contents quickly becoming outdated, another problem with logbooks was that they only described a single voyage, limiting the spatial scope of the information. An attempt to gain a firmer grip on the phenomenon was made through the publication of tables. Once more, the failure of early versions to attach dates to each triplet of coordinates rendered them unfit for mariners in following decades to steer by. A table submitted to press by Petrus Herigonus in 1644 does, however, clearly evince the intended practical application at sea, all locations being neatly ordered by longitude. De Groot and Vooght (1684) explicitly stated that their compilation was to be used to estimate longitude, but also warned of time-dependent mutations. Other tabular examples include navigation textbooks by de Graaf (1658), de Decker (1659), van Nierop (1676), and Vooght (1706). Mathematical practitioner J. Pietersz even advertised a separate table for sale in 1699, written in Dutch and French.[31] Unfortunately, the quality of the presented findings was rather variable, sources often remained undisclosed, and rapid alteration of declination in some regions severely limited their temporal window of application.

Emerging Contours

Once magnetic data had been compiled, various ways existed of processing the information. One could list them in tables, for example ordered by latitude, longitude, or geographical region, as was done in some navigation textbooks. Alternatively, they could become part of sailing directions, as one of several types of hallmarks associated with specific locations. In order to get a better sense of the spatial distribution of declination, however, a visual representation on a map or chart worked best. In its most basic form, a wind-rose oriented away from geographic north could convey the notion that the whole displayed area was subject to the same needle displacement (an obvious simplification). The idea to plot a series of readings, for instance obtained during an ocean traverse, seems to have emerged in the late sixteenth and early seventeenth century. On his 1576 voyage of discovery, Martin Frobisher used a chart specially prepared by William Borough to mark magnetic observations with tiny arrows. In France, Jean Guérard of Dieppe used a five-degree grid to position his data on a Mercator projection of the Atlantic, while in Spain Diego Ramírez de Arellano adjoined a chart to his printed description of a voyage to the Magellan Straits (1620) to illustrate the gathered declination data. Robert Dudley's Florentine sea atlas *Arcano del Mare* (1646–47) featured no fewer than 127 charts on which he annotated local compass behavior at various places. Even as late as 1788, French Academic Le Clerc (better known as the Count of Buffon) included seven "Magnetic Charts" in his *Natural History of Minerals,* showing plotted declinations and inclinations.[32]

The decision to connect all points of equal value with an unbroken line heralded the birth of magnetic thematic mapping. The introduction of *isolines* constitutes a fundamental rethink, visualizing the field as a global phenomenon rather than a number of scattered point values. Isolines provide a substantial abstraction from individual measurements by forming closed curves. They demarcate the boundary between two mutually exclusive regions, and outline the shape and extent of maxima, minima, and intermediate levels of the explored property, combining both data reduction and interpolation. This is of particular interest in the geomagnetic case, since the graphical delineation of a prior supposition of regularity works equally well without supporting empirical evidence. The earliest isogonic maps actually served exclusively to highlight imposed rather than observed order.

Technically speaking, it was Alonso de Santa Cruz who was the first (in the late 1530s) to rely on isogonics to visualize his geomagnetic hypothesis. It will be recalled that his longitude system worked from the assumption of a tilted dipole, with a single great circle agonic, and maximum declination at ninety degrees distance there from. Every fifteen degrees of displacement east and west would result in a change in needle deflection of half a point. When the Spanish king ordered his cosmographer to elaborate upon the global distribution of the Earth's magnetic field, the latter had a world map drawn (now lost), to which he merely added the predicted declination underneath each meridian passing through fifteen degrees of longitude or a multiple thereof. These meridians thus doubled as isogonics, partitioning the globe in a regular pattern. De Santa Cruz neither plotted any data, nor averaged values per grid cell, nor traced complicated curves. A few decades later (1584), Peter Bruinsz in the

Low Countries drew the first true isolines; these were isobaths, lines of equal water depth.[33]

It was Italian Jesuit and navigation teacher Christovao Bruno who, in the 1620s, became the first person to draft nonmeridional isogonics for the Atlantic and Indian Oceans. Although this chart no longer exists either, two near-contemporary sources provide enough descriptive details to assert the inventor's primacy in this respect. One of these is Kircher's *Magnes* (1641), which itself partly relied on an earlier report from a fellow Jesuit in Madrid. Bruno had made several sea journeys to India, which may have exposed him to relevant roteiros, and enabled him to make his own observations. He introduced the concept of isogonics for each degree of declination. Kircher remarked that Bruno's lines had run parallel, something he considered impossible; a true representation would produce not only straight lines but also curved and broken ones, he figured.[34] Intriguingly, a manuscript written by the inventor himself, which explained the method for voyaging to the subcontinent, stated that the Atlantic agonic did not cross the equator perpendicularly (which was indeed the case at the time; see figs. 3.1 and 4.1). It suggests that both Bruno's hypothesis and his chart were not solely founded on theoretical constructs of meridional agonics. In addition, any reasonable alignment of isogonics at a sufficient angle with the horizontal would in principle have been adequate (in combination with observed latitude) for determining longitude.[35]

Someone who was certainly aware of this fact was Edmond Halley. He had read Kircher's work, and had doubtless seen contemporary thematic charts of ocean currents. He may also have been inspired by isobaths, or his own 1686 chart of wind systems, in which he had used strokes with and without arrowheads to indicate direction. The astronomer produced three isogonic charts around the turn of the eighteenth century. The first was a polar projection of the Antarctic in manuscript (1695, now lost), presented to the Royal Society to illustrate "the several variations, exhibiting at one view the several tracts wherein the variations of the magnetical needle are regularly east and west."[36] This attempt was clearly longitude related, and since at the time precious few observers had performed magnetic measurements at extreme southerly latitudes, it seems likely that the main sources for the drawing had been Halley's two south poles as envisaged in his 1683 hypothesis (see fig. 3.3).

Halley's other two isogonic charts appeared in printed form, and covered respectively the Atlantic (1701) and the whole world (1702). Both were published by the newly established firm of Richard Mount and Thomas Page of Tower Hill in London. In the accompanying description, Halley claimed the idea of isogonics as his own, calling them "curve-lines." This was an awkward choice of words, since the same term then generally denoted the pattern displayed by iron filings around a magnet; the confusion could easily lead to mistaking lines of equal angle from true north for the direction of magnetic force. Nevertheless, the widespread distribution of the two published charts throughout western Europe in the ensuing decades ensured the rapid popularization of the isoline concept. The charts were founded upon the dataset that Halley had personally compiled at sea in 1698–1700. In addition to the Atlantic isogonics of the first printed edition, the 1702 *Sea Chart of the Whole World* had isogonics traversing the Indian Ocean, based on several (presumably recent) logbooks by others. The continents and the entire Pacific basin remained geomagnetically blank.[37]

Halley did not simply produce the charts as a lasting record of the surveys, but intended them for practical purposes as well, in particular at those places where isogonics ran roughly north-south and close together. Theoretically, such a configuration would enable a mariner to estimate his position along a parallel, a partial and localized solution with which to find longitude, for as long as the chart remained accurate. This restriction is quite important; the astronomer took care to stress that his work was only a snapshot of the situation in 1700, and would require future amendments to remain up-to-date. In a preemptive gesture, he supplied his audience with predictions of (regular) change at a few locations, attempting to keep the contents of his charts fresh for a little while longer.[38]

Halley's isolines kept quite a few pens in motion over a period of decades. A favorable reception in England and the Dutch Republic contrasted with predominantly negative reactions in French circles. In the *Mémoires* of the Paris Académie de Sciences over the years 1701–33, the subject came up on nine occasions. The Frenchman best equipped to assess the merits of the charts was Guillaume Delisle, who had compiled a dataset two orders of magnitude larger than Halley's. He submitted his views to the Académie's journal in 1710, focusing on the most obvious weakness of the whole enterprise: the unpredictable nature of secular variation. With examples from European and Atlantic stations, he underlined that isogonics did not simply travel in longitude, but changed shape and orientation as well. Having made an extensive study of secular acceleration and the inconsistency of westward drift, he could have listed many more instances than he did. Halley's facile reaction came five years later in the *Philosophical Transactions*. Instead of addressing this fundamental issue, he merely commented on a few individual locations where his chart differed from recent observations. He did admit that some slight inaccuracies could have resulted from extrapolation for parts he had not visited himself, but he maintained confidence in the project as a whole.[39]

In the decades following the initial publication of Halley's isogonic charts, various reprints appeared in England and on the Continent. The first to completely fill the Pacific gap was Dutch scholar van Musschenbroeck, in his 1729 Ph.D. dissertation, postulating a third agonic there. Alternative charts were moreover constructed by French engineer Frezier in 1717, and former VOC captain Nicolaas van Ewyk in 1752. Ten years earlier (around 1742), the proprietors of Halley's printed charts, Mount and Page, had decided that an update of his work was long overdue. Accordingly, they employed Charles Leadbetter, a teacher of, and writer on, astronomy, mathematics, and navigation. He was to process a new dataset of circa 1,100 observations compiled by his colleague Robert Douglas, who had taught navigation on board Royal Navy ships. Unfortunately, Leadbetter's method of reducing information to a chart relied not so much on the calculation of spatial averages as on the linear extrapolation of past trends of local secular change several decades ahead. As Delisle had already pointed out, this was an inappropriate technique for evaluating geomagnetic data, and the attempt soon ended in failure.[40] This left the Tower Hill stationers short of both a reliable chart and a capable successor to Halley. But in 1744 their fortunes changed for the better, when they enlisted the services of Royal Society Fellows William Mountaine and James Dodson, both teachers of mathematics. Their partnership would last until Dodson's demise, after which Mountaine continued to revise some of the publishers' charts and nautical textbooks for many years on his own.

The task the pair set themselves in 1744 was far more ambitious than Leadbetter's halfhearted effort had been, asking and gaining permission to examine the masters' logbooks of the navy, the EIC, and the Royal African Company. For unknown reasons, the Hudson's Bay Company refused a similar request. But additional data from individual voyages helped to further strengthen spatial coverage, based upon Douglas's set, East Indian measurements from William Jones, Anson's circumnavigation of the globe in 1740–44, and Middleton's Arctic exploits. Nevertheless, certain regions remained underrepresented in the sample; some isogonics were consequently drawn as broken lines where data were sparse. The chart, when eventually published in 1745, was no commercial success, but did receive good reviews from scholars.[41]

Recognition of the field's continual change espoused the need for a second revision a little over a decade later. In 1756, work started on a data compilation that was to dwarf all previous attempts in scope. This time around, the two mathematicians initially hoped they could build upon their earlier work, reconstructing the paths of isolines by analogy, based upon principles of regular change yet to be established. For this purpose, they divided the most frequented oceans up in a five-degree grid, and tracked the declination values within each cell in 1700 (based on Halley's charts), 1710, 1720, 1730, 1744, and 1756. They furthermore supplemented the existing dataset with records from Royal Astronomer Bradley, the Hudson's Bay Company (this time compliant with a renewed request), and more material from recent East India Company and navy voyages, reaching a total of over fifty thousand observations. Tabulated preliminary results appeared in the *Philosophical Transactions* for 1757; these reasserted once more the futility of searching for fixed rules regulating time-dependent variation. Thus Mountaine and Dodson regretfully concluded: "We have been obliged to pursue our former tedious, but more safe and justifiable method of proceeding, which was by collecting the greatest number of observations possible."[42] Sadly, Dodson did not live to see the end result; he died before 13 April 1758, when Mountaine finally presented the new isogonic world chart for the year 1756 to the Royal Society, together with a pamphlet explaining its application. He envisaged a future third revision as well, but this project never came to fruition.[43]

The 1756 chart was sold at Amsterdam in both a Dutch and a French translation around 1760, while another version printed in France (with identical isogonics) reckoned the longitudes from the Paris meridian. Five years later, Jacques-Nicolas Bellin, naval engineer, geographer, and author of *Neptune François* (1753), included the work in his *Petit Atlas Maritime*, but with important alterations. Not only did he print the whole on a smaller scale and place the longitudinal zero at the French capital, he also substantially modernized the geographical base map (unchanged since Halley's days in both English revisions). The isogonics themselves he left untouched; in order to make them reflect the situation of 1765, Bellin assumed a uniform westward shift of between nine and ten minutes annually. Instead of having to redraw the whole, the cartographer opted for a shortcut, advising his readership to simply add one and a half degree of northwesting to the copied values.[44] In the last quarter of the eighteenth century, new isogonic charts of precious little fame occasionally saw the light of day, produced by mathematical practitioner Samuel Dunn (1775, Atlantic and Indian Oceans), natural philosopher Johann Heinrich Lambert (1777, world), French scholar Le Monnier (1778, world), and major James Rennell (1798, Africa). The isogonic globe and *Magnetic Atlas* by American surveyor John Churchman have already been mentioned.

Isogonic Navigation?

All this activity in charting declination naturally leads to the question, For whom were these efforts intended? Did isoline representations of the geomagnetic field actually reach the navigator? Were these charts ever supplied by maritime organizations as standard issue? Did the helmsman rely on them, rather than on values obtained through observation, to correct his steering? And did they constitute a trusted means to finding longitude at sea? In fact, very little evidence of any practical implementation at sea exists. There are, on the contrary, a number of substantial indications, rooted both in geophysics and primary sources, supporting a negative answer to these questions.

As far as navigational practice is concerned, a few of the publications listed above can immediately be ruled out: the attempts by Frezier, Musschenbroeck, Lambert, Le Monnier, and Rennell appear primarily intended as illustration, while Frezier's, Le Monnier's, and Churchman's drafts probably also served to draw imposed regularity rather than actual data. Similar question marks rise regarding the empirical value of Dunn's work, which was additionally drawn on an unorthodox projection. Van Ewyk, on the other hand, had most likely based his work on many measurements, but it suffered from a lack of data abstraction (resulting in highly irregular curves), and also relied upon an unsuitable scale and projection for practical use. Mountaine and Dodson's 1744 chart similarly had a very limited circulation. This leaves the two publications of Halley, Mountaine and Dodson's 1756 revision, and Bellin's 1765 French edition.

Halley's Atlantic chart was usually published on a single sheet of about 22 by 20 inches, the more elongated world chart on a double sheet with combined dimensions of roughly 57 by 20 inches. From this surface area, border margins, dedication, and an extensive description still had to be subtracted. Mountaine and Dodson's two versions of the world chart were of similar size, and Bellin's 1765 issue was drawn on an even smaller scale. In the instruction accompanying his own 1775 isogonic representation, Dunn observed: "The smallness of the scale to which all those former charts have been drawn, could not but render them very unfit for any attempt of the longitude at sea by them, however correctly they may be supposed to have been drawn."[45] Indeed, Halley himself had acknowledged in 1701 that his chart was wholly unsuited for coastal navigation, on account of its small scale. Another comment, applicable to the charts by Halley and Mountaine and Dodson, is the reliance on the same geographical base map (made by Halley). Around the turn of the eighteenth century it was probably a little better than some of its ordinary Mercator competitors. However, half a century later geography had naturally made many new and important advances, which failed to be reflected in Mountaine and Dodson's work. Later competitors did not fail to exploit this weakness to highlight the supposed higher quality of their own wares.[46]

But fellow cartographers can hardly be considered unbiased, having their own commercial interests at heart. Unfortunately, when one examines contemporary navigation manuals, opinions are found to be even more critical. A massive, classic Dutch textbook by Gietermaker (1757) devoted merely a single sentence to one of Halley's charts, used only to illustrate the spatial variability of the field, and without hinting at any kind of practical application at sea. French examiner Bézout, in 1775, questioned the production process of isogonic charts in general, on account of magnetic observations being both insufficient in quality and quantity to allow the confirmation of any values. Dutch captain J. O. Vaillant, who published a treatise on navigation in 1784, found both Halley's and Mountaine and Dodson's attempts "very defective." Mathematical practitioner and astronomer Lasalle (in 1787) was more categorical: "All charts of this kind . . . are in their perfection still well below what they should be."[47]

The fundamental point of criticism was a geophysical one: the field was altering all the time, at different rates at different places, and the rates of changes themselves were subject to variability as well. This prevented any linear extrapolation of a recorded progression into the future. All previously mentioned authors made this insoluble problem central to their argument. Amsterdam lecturer and examiner of masters Pybo Steenstra in 1770, moreover, reminded his audience that tables such as those published in 1757 by Mountaine and Dodson would have to be updated every three to four years to remain of use. A little earlier, Bouguer, writing on the 1744 edition, also noted that mariners hardly displayed any interest in such charts.[48] This was not surprising; modern time-dependent geomagnetic models illustrate that all geomagnetic charts should have had a label informing the customer of their expiry date. For the Atlantic, and especially for the Indian Ocean, this date lay only a few years after publication.

A closer look at the surmised practices on board ship is clearly necessary. The assumption that navigators corrected their ships' headings by consulting isogonic charts made by natural philosophers, rather than trusting their own experience and direct astronomical observations, does not seem very likely, for a number of reasons. To start with, there is the temporal point to consider: except for a few years immediately following data compilation, all isogonic charts were seriously out of date, and consequently dangerously incorrect on numerous locations. Both Halley's world chart and Mountaine and Dodson's 1756 revision were already at least two years behind the times when they finally came off the press. Modern comparison of these charts with a contemporary geomagnetic field model reveals significant differences after less than a decade.

Secondly, is it realistic to expect a seasoned navigator to simply give up an entire source of available information, which also happens to be accessible through one of the least complicated measurements in ocean navigation? Would he confidently bet the safety of cargo and crew on a single chart, when a few minutes' work would yield a current fix accurate to less than half a degree? Imagine for a moment that he would, and that he eventually reached a place where the value estimated from the chart was actually off by several degrees. Not only would he make an adjustment in steering, but also in the dead-reckoning, enlarging the positional error. The following day, the chances of obtaining a compass correction more divergent from the true needle deflection would therefore in all probability increase, leading to an even bigger discrepancy between reckoned and actual position, and so forth. Only a very foolhardy, stubborn individual would maintain such unwavering confidence for long, especially in the presence of other warnings (such as observed latitude) that something was seriously amiss. The dangers of ocean travel furthermore wield the principles of natural selection against such behavior.

Even more poignant is the absence of a multitude of such errors in logbooks. As previously discussed, the navigator observed his latitude whenever possible, for comparison with his dead-reckoned estimates. Additionally, he eagerly interpreted the most tentative clues from the natural surroundings (water color, birds, seaweed), while noticing compass deviation of a few degrees, and identifying the cause. All these instances appear in the logs. Yet one is apparently to assume that navigators would refrain from giving their opinion on isogonic charts, whereas other charts taken along received criticism and corrections on numerous occasions; it seems an improbable scenario. Masters and mates routinely noted the discrepancies they found between charted positions of landmarks and observed coordinates, and present field reconstructions certify that substantial differences existed between measured and depicted declination for many locations throughout much of the eighteenth century. It would surely be very remarkable if seafarers had refrained from giving their opinion concerning predicted declination, if they *did* keep such charts on board. Indeed, sailing directions explicitly encouraged ships' officers to correct reported coordinates, magnetic declinations, and other relevant information, where found at odds with observation. What possible reason could warrant such an exclusive veil of secrecy to shroud isogonic representations, in this otherwise so practical profession?

Positive evidence is similarly lacking from ships' inventories. Given the fact that Halley's charts appeared in print in the first two years of the eighteenth century, and that they quickly developed an ever growing discrepancy with reality, one would expect an official introduction to have taken place sooner rather than later. Unfortunately, only sources from the Dutch East India Company offer a chance to investigate this matter. This rather bureaucratic organization kept lists of standard-issue charts for each and every ship of consequence departing for an ocean crossing, both at home and at Batavia (Java). The strategically valuable material had to be signed for upon receipt (at departure) and return (upon arrival at the destination). In the East, the navigators had to make do with local paper charts, while the parchment Mercators remained in harbor under lock and key until such time as a homeward-bound crew required them again. Fortunately, some of the company's Asian checklists have survived. A sample of eighty-five over the period 1688–1719, studied in search of any notable change after 1701, has shown that the basic set remained completely unaltered over the examined interval. More significantly, not a single reference was ever made to any isogonic chart. Subsequent introduction in the 1720s or 1730s seems unlikely on account of secular variation, which practically rules out useful application of Halley's magnetic information.

Mountaine and Dodson's exploits, however, temporarily reopened the window of opportunity. Around the same time, the preponderance of Dutch sailing directions exhibited a marked growth; over the ensuing decades the Dutch East India Company compiled, edited, and issued dozens of versions. The

Earth's magnetic features formed an important component of many such instructions. Sailing directions clearly illustrate a heightened awareness of secular variation, and bear witness to usage of declination as a positional indicator, for example: to establish the proper longitude at which to cross the equator, the distance to the Cape of Good Hope, and the nearness of certain islands and other way stations. If isogonic charts had been carried on board for longitude purposes, it is in these documents that one would expect to find references to them. The sources are, however, unanimous in maintaining a deafening silence on the matter. Cited values of declination appear to stem solely from the study of recent logbooks, handed in upon completion of a voyage. New editions of the same text, with only the declination figures altered, again testify to awareness of secular change, seriously at odds with reliance upon isogonic representations of decades ago.[49]

Lastly, a total of 1,630 eighteenth-century logbooks from several nations has been examined in search of evidence. Rather surprisingly, French material did yield a single certain reference to an isogonic chart. It unequivocally identifies Bellin's 1765 edition, and explicitly links magnetic observations with longitude determined from that chart. It was made by an unknown officer of the French Navy's *Fendant*, part of a squadron on its way to Mauritius. The year was 1782, so the chart was already well into its second decade. Little wonder, then, that dead-reckoned and "magnetic" longitude differed more than marginally: on 23 April by 8°55′, three days later 6°45′, and the day after 6°03′. The navigator was not amused, and wise enough to keep to the standard method of dead-reckoning.[50] Of course, this single instance carries no statistical significance whatsoever; in the absence of further evidence, it seems safe to conclude that isogonic charts did not play any substantial role in practical navigation in the eighteenth century.

Proportional Representation

On 16 May 1666, Henry Phillippes, author and editor of several works on navigation, presented a paper at the Royal Society, accompanied by a chart in folio, which was later bound in the "Classified Papers" of that institution. Thus preserved for posterity, it constitutes the oldest extant representation of multiple ocean voyages to detect an underlying regularity in observed magnetic declination. On a plane grid, the outlines of the African continent and several East Asian coasts and islands have been sketched, with some notable landmarks identified. Most of the northern hemisphere contains text captions, which identify a 1607 journey by John Davis and a 1641 venture by William Mynors as the two course plots prominently crossing the southern hemisphere. All measurement points are connected dots, with the values of declination noted adjacent to them. The route chosen by Davis was more northerly than that of Mynors, the former stopping over at Madagascar.

The most intriguing part of the manuscript consists of a table constructed by Mynors, oriented horizontally in order to more or less align longitudinal entries with their position on the chart. Phillippes explained: "Captaine W.m Mynors finding . . . that, in sayling somewhat southward of the parallel of the Cape, the needle did alter a degree of variation about every 100 miles of east or west, did accordingly frame this table for his owne use, whereby he know more certainely how far he was from the Cape then by any reckoning he could keep any other way."[51] The idea "to proportion the variation to the east and west of the Cape, either to the longitude or the distance in miles or leagues" represents another conceptual leap in the processing of magnetic data. The assumption of a regular progression of magnetic declination as a function of easting and westing between distant locations has occupied many minds as a potential method to determine longitude. At the basis lay the assumption that isogonic lines would be evenly spaced (but not necessarily meridional) over a certain area or trajectory. As in Phillippes's case, a limited number of observations provided the foundation. The concept falls short of being classified as a geomagnetic hypothesis, because coverage was never global, and did not imply an underlying dipole or multipole system. Nevertheless, the attempted pattern recognition would work only at the expense of a substantial reduction of the available information. Part of the problem was the separation of signal and noise: by allowing a larger range for instrumental and positional error than is presently assumed to have hampered historical measurements, readings became more malleable, supporting interpretations of regularity. Moore's navigational textbook (1681) stated that "it's regular and orderly to them that well observe it," and blamed sloppy observational procedure for false notions about irregular field characteristics being spread. Harrison, just before the turn of the eighteenth century, believed the method to be applicable over limited areas: "The difference of the variation between any two places in East-India may be always near the same; and not only there, but in most other parts of the world . . . that are not above 20 or 30 degrees distant asunder."[52] French hydrographer Guillaume Delisle, a few years later, with the aid of a vast collection of magnetic

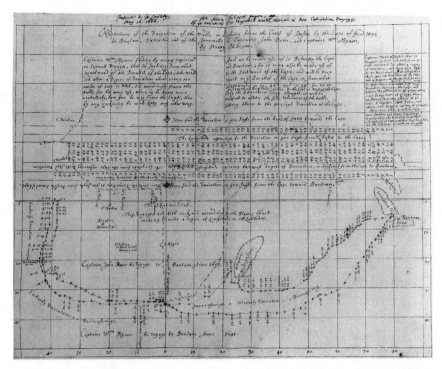

Fig. 7.1. The Phillippes chart (1666, in RS, CP 9 [2] no. 2 fol. 195–96). Copyright ©
Royal Society, London.

observations, equally predicted intervals of regularity for some ships' tracks,
for example between the islands of Réunion and Rodrigues.[53]

Of all locations put forward as displaying some order in the horizontal
magnetic vector, the Cape of Good Hope was far and away the favorite. In the
description of his 1701 Atlantic isogonic chart, Edmond Halley had remarked
upon the almost meridional orientation of closely spaced isogonics near
southern Africa, and their potential use for estimating longitude. D'Après de
Mannevillette reiterated the point in 1765, stating that "these declinations appear to keep between them such a regularity when one travels from the west to
the east, or from the east to the west, that one can consider them as means to
become aware of . . . errors in the reckoning."[54] The author furthermore employed a perceived failure to recognize such underlying trends to chastise in-
difference on the part of navigators, in relying on outdated sources and not
seizing all opportunities to observe.

The above-mentioned attempts to detect regularity can be interpreted either as an extreme form of processing empirical data, or as the least far-reaching extrapolation of observed characteristics into theoretical constructs. The degree of complexity inherent in any spatial representation of the Earth's magnetic field was to a large degree reliant upon the attributed error margins of the magnetic measurements that founded them. If the observational error was deemed large, then the resulting reconstruction would display little complexity, and a limited set of parameters could adequately represent them. The master, on the other hand, found himself at the other end of the spectrum; he had to deal with the rough reality of a surface distribution that defied such imposed simplistic patterns. The practical requirements of seafaring denied him access to the level of abstraction attained by natural philosophers and longitude-finders. Instead, navigators had to deal with field peculiarities on a day-to-day basis, both as an obstacle to be overcome and as a potential positional clue. Doubtless, they too had a mental image of what declination to expect where, but it was patchy and did not conform to a small set of fixed rules predicting the field's shape and change everywhere.

This leads to the intriguing question of how well hydrographers and seafarers have been able to deal with this spatially and temporally variable quantity. Hundreds of predictions, recorded in sailing directions over the years, detailed how much magnetic declination to expect at a certain place and time. In addition, many thousands of steering corrections, preserved in the logbooks of several nations, offer a momentary glimpse into each mariner's expectations regarding how much declination to allow for. When considered individually, these notes carry no information beyond their immediate context. But in large quantities, and grouped per country, traversed oceanic region, maritime organization, or time period, they bear the distinct hallmark of a certain level of accuracy. With an independent, geophysically constrained model of the field's behavior in former times, this aspect of early-modern navigational practice can now be quantified and analyzed for the very first time. This novel approach and its rewards are the subject of the final chapter.

Quantifying Geomagnetic Navigation

The bond between geomagnetism and matters maritime is of a quite venerable age. From the early days of oceanic navigation, mariners had to deal with the vagaries of the needle on a daily basis, observing, estimating, and compensating for the field's peculiar features wherever their ships took them. An incessant stream of empirical data resulted, carried for future reference by the navigators themselves, and used by hydrographers to draft and update sailing directions. These two professions were thus mutually dependent upon one another. In a way, the current geomagnetic research mimics this symbiosis: in both instances, logbook data offered the possibility to track the field's secular change. The results were in the past applied to improving navigational accuracy, and in the present to studying it. In each case, the exchange was beneficial to both parties concerned.

The following analysis relies upon a modern geophysical reconstruction of the Earth's magnetic field. A time-dependent model at its origin, deep inside the Earth, is the most effective means to analyzing both global and local patterns in the historical past. The model offers a physically constrained description of the field over the interval 1590–1800. It shows how some features have remained steady over time, while others changed shape and migrated, the waxing and waning magnetic shadows of the convecting molten iron in the outer core. No previous attempt, either by geophysicists or historians, has been able to monitor historical secular variation this far back in time.

Ideally, the data should describe the field everywhere and continuously, recording both magnitude and direction of the geomagnetic vector with perfect instruments. Nowadays, it is possible to obtain planetary coverage at high resolution with the aid of satellites. We can usually put them in the proper orbit, but we cannot send them back in time. A numerical approximation of past developments based upon historical observations is the best alternative. Individual compass measurements yield a mere snapshot of local conditions, and contemporaneous published sources render only the vaguest of outlines. Even consistent series of readings made at a fixed location do not tell much; the "hidden level" of global secular variation patterns would still remain largely undisclosed. But using tens of thousands of data points from all over the world, with information on where, when, and how accurately they were obtained, in combination with several physical assumptions concerning Earth's inner structure, it is possible to build a carefully weighted estimate of how the field has evolved over the last four centuries. Unfortunately, both the number of measurements and the amount of detail tends to decrease further back in time, a problem insoluble by any amount of mathematical tinkering. A stupendous quantity of potentially valuable material from these early times is forever lost. It is therefore of paramount importance to save and savor what is left, and this requires the sifting of thousands of sources, to obtain the best possible result. The required dataset for reaching this goal lies buried in the maritime historical heritage of many countries.

The reason to delve into navigation as practiced on board ocean-going vessels from the late sixteenth to the eighteenth century will be clear: navigational logbooks carry the most direct evidence of compass use on board ship, providing the data for the model. These documents also contain information on the types of instruments and measurements, the frequency and accuracy of the results, and the way seafarers daily dealt with the field's changing appearance. In addition, such data provide insight into the error margins associated with particular observations. Other sources (sailing directions, navigation textbooks) can further elucidate the important place that magnetic declination held in the minds of those who had to negotiate oceans. Aboard ship, it directly affected steering, dead-reckoning, charting, cross-bearings, and estimations of leeway. In addition, maritime sources can also illuminate the considerable extent of the mariner's reliance on geomagnetism for determining his position at sea. Given its importance in practical navigation, the question then naturally arises as to how well people in the past have been able to deal with this unpredictable phenomenon. This chapter will attempt to establish the extent to which a quantified geophysical assessment can answer this question.

The ultimate goal of the current research was to establish the benefits of mutual exchange between geomagnetism and history. It linked the two disciplines by comparing a geomagnetic model with historical estimates of local magnetic declination. The purpose was to find marked differences in accuracy between nations, periods, traversed oceanic regions, and maritime organizations. The geophysical model was thus fed back into the field of historical inquiry, uncovering a layer of information that has remained largely inaccessible

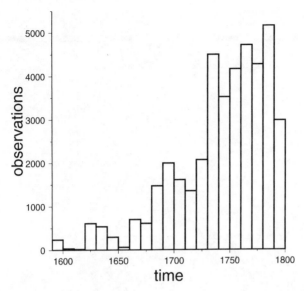

Fig. C.1. Histogram of the total number of gathered observations of magnetic declination per decade. N = 51,306.

through regular historical studies, based on more limited numbers of documents. The processing of a large number of sources not only served to construct a robust field approximation, but also offered a chance to quantify group characteristics of historically relevant entities.

Geomagnetic reconstruction and historical estimation thus each provided one side of the equation. This strategy embodied the necessary and sufficient steps to test the validity of the approach, in particular whether it could reap additional benefits relative to more traditional modes of historical research. To that effect, a quantitative comparison of modeled and estimated declination in the maritime realm was to encompass the efforts of hydrographers and navigators. Positive results in any one of these two areas would suffice to vindicate the interdisciplinary communication.

The fundamental idea was thus to determine whether the time-dependent model could be used as a touchstone for examining the historical past in a novel way, or, phrased in the form of an elementary question: Does a geophysical reconstruction of geomagnetic secular variation allow the answering of historical questions that traditional approaches to maritime history are unable to address? This question was posed in such a manner that a plain yes or no

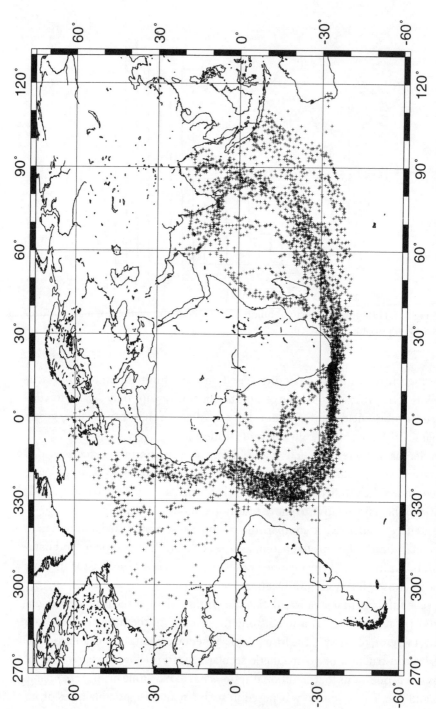

Fig. C.2. Collected maritime geomagnetic observations from the period 1651–1700, plotted on a cylindrical-equidistant world chart. Shipping lanes are clearly delineated, due to the prevailing winds and currents. $N = 5,700$.

would suffice, and the answer will become apparent in the course of the remainder of this chapter. But before establishing to what extent the cross-fertilization bore fruits, the Earth's magnetic behavior in time and space first needed to be properly reconstructed. Following is a brief description of how the geomagnetic model came into being.

Data Processing

The collaboration between history and geomagnetism initially took shape in joint data collection. The shared intentions were threefold:

1. To exploit the vast amounts of historical magnetic data available in archives and museums, so as to improve the accuracy of field models.
2. To push the temporal boundary as far back as sources would permit.
3. To properly quantify the errors affecting these data.

The geomagnetic model used here relied almost entirely upon original maritime declination measurements from the period 1590–1800, augmented with a small set of 156 inclination observations. In addition to locating and processing sources that contained geomagnetic data, the historical research made an in-depth survey of the measurement practice, regarding its frequency, types of observation, instrumental accuracy, potential compass defects, and other impediments. It also examined the problem of quantifying the longitudinal mislocation in oceanic dead-reckoning as a function of time. The substantial sample of ships' logbooks used in this analysis has brought several characteristics to the fore. The following paragraphs will briefly survey the process from manuscript to model.

A total of 2,062 logbooks was located and processed by the author, in various Dutch, French, English, and Danish repositories, over a period of about three years. In all, 1,339 logs from the surveyed countries contained useful information for geomagnetic modeling, resulting after dismissing doubtful points in 51,306 observations of magnetic declination. Longitudes reckoned from fifty-eight different meridians required recalculation relative to Greenwich. East India companies proved to be the most productive shipping category by far. Tables A.1 and A.2 in the appendix give some indication of their yield compared to that of navies and smaller maritime organizations such as private merchants, slavers, and whalers. The associated number of logbooks precedes the size of data subsets.

In order to derive the best possible model of the Earth's magnetic field as it has evolved through the ages, each data point's fidelity had to be assessed. The modeling algorithm had to properly account for the main factors affecting magnetic readings by assigning each a weighting that expressed its relative reliability. Accurate measurements, made at well-defined locations, should dominate over less certain readings, made after months at sea without geographical confirmation of position. The calculations also incorporated existing geophysical assumptions concerning crustal magnetic fields, which are considered a source of noise when imaging the core field.

How is one to factor in the errors associated with this historical maritime dataset? Fortunately, the gathered observations from logbooks constituted a vast resource of previously untapped information on positional and observational accuracy, promising a far more accurate model than obtainable by relying on unweighted data. The quantification of these factors was therefore divided into a positional, and an observational phase. The first started with the separation of the data into subsets of twenty and fifty years and plotting each of these on a plane world chart (see fig. C.2 for an example). These early plots showed that some typographical errors had eluded detection in the archives: a few ships' tracks still ploughed through land. In several cases, this was simply caused by incertitude regarding the actual longitude of mentioned places while processing the data. Additionally, in English navigational practice, a land sighting would often imply a new meridian to reckon from, so a whole *voyage leg*, that is, a sequence of points between two land sightings, could be artificially translated hundreds of miles from its true position. Both situations were easily recognized. Identifying the right location on a chart proved more troublesome at times; on numerous occasions, it required comparing a contemporary chart containing old nomenclature, the course plot, the logbook reference, and a modern atlas.

A more insidious problem regarding navigational errors was the positional incertitude during ocean traverses, prior to the advent of reliable methods with which to determine longitude. Because the only viable alternative to the mariner was the daily dead-reckoning, navigational errors are often serially correlated; an estimate of longitude on a given day was obtained by adding an approximate change to the longitudinal estimate from the previous day. Positional error therefore tended to build up, the longer a vessel remained out of sight of land. Only when a familiar coast was actually sighted could the reckoning be restarted, and the longitudinal error be reset to zero. Mariners usually knew the coordinates of their ports of departure and destination, and tended to interrupt extensive voyages by one or more intermediate land sightings, partly to correct their reckoning. Each positional fix now allows comparison of a known geographical location with the dead-reckoned one, yielding an estimate of the average accumulated longitudinal error along a voyage leg.

Each leg had therefore to be separately assessed. This procedure was one of the many tasks performed with the aid of the purpose-built *voyage editor,* an interactive graphical user interface. Its main window offered a view of the world in four magnifications. This enabled the user to zoom in on specific regions and examine voyage data in relation to nearby shores and islands. Its main function was to correct longitudinal error accumulated between successive landfalls. The program allowed single-point editing (1,082 instances), as well as translation (635 points) and stretching of voyage legs (29,954 points). These latter two operations acted on multiple points: the user first chose a start and end point to define the voyage leg boundaries, and could then pick up either end of this sequence and transport it elsewhere. In the case of translation, this displacement would affect all points uniformly, shifting the whole leg to the east or west, while maintaining the recorded latitudes. Alternatively, in the case of stretching, one end of a leg remained fixed, while the other distances between points became longitudinally "stretched" to meet the translated other end.

For example, when mariners traveling eastward from Tristan da Cunha (in the South Atlantic) finally spotted the Cape of Good Hope, they may have severely under- or overestimated the westerly winds and currents. The reckoning would consequently put the vessel either far to the east or to the west of Cape Town. The last land previously sighted (the islands of Tristan) could then be fixed as a starting point in the voyage editor, and the logbook entry containing the comment "seen the Cape of Good Hope" or "at anchor in Table Bay" moved to the appropriate location on the chart. The program would subsequently recalculate all longitudes to reflect the change proportionally over time. After examination of typographical errors, meridians, and internal consistency, followed by translation, stretching, and analysis of distance traveled in the voyage editor, a total of 136 points still remained implausible, and were consequently discarded. Of the others, both the original and the corrected longitudes were kept on file, offering the possibility of quantifying positional uncertainty later on.

This required a theoretical framework that classified voyage legs into one of two possible categories. The first was a leg that started from a known position, but with an unverifiable endpoint, somewhere at sea; the log would then con-

Fig. C.3. An East In-
diaman's outward and
homeward journey in
1743–45, plotted by the
voyage editor in lowest
magnification. Points
represent compass ob-
servations and land
sightings only.

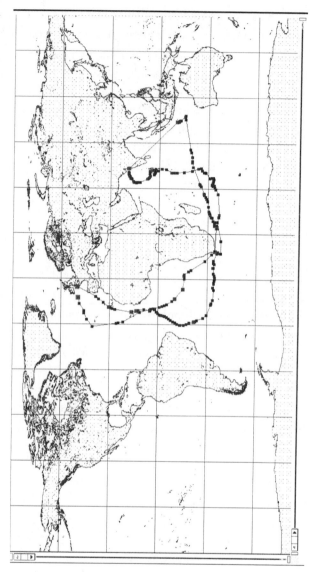

tain a number of observations before it finished without a geographical fix, and only dead-reckoned longitudes would be recorded. This type of movement is generally called a *random walk* model, also known as *Brownian motion*. The second category comprised those voyage legs with both initial and final positions known. The statistical literature refers to such a sequence as a *Brownian bridge*.[1] The data from each leg were in this case stretched in the voyage editor (see above), annihilating the positional error at the end point by bringing the dead-reckoned longitude into agreement with the coordinates of the identified landfall. The remaining positional uncertainty consequently reaches a maximum at the temporal midpoint of the journey, while it is zero at both extremities. Furthermore, the size of the maximum error of the Brownian bridge is only half that of an ordinary random walk of the same length. The points making up a leg between two known geographical locations are thus positionally less uncertain than those of an open-ended ocean journey, as would be expected.[2] Furthermore, the longer a vessel remained at sea without geographical confirmation of position, the larger the accumulated error in longitude would become. By using the information obtained while correcting landfalls in the voyage editor, it became possible to quantify the ship's margin of mislocation as a function of time. The applied approximation assumed an initial reckoning error of about 0.4 (decimal) degrees to grow with the square root of the number of days since the last land sighting. Thus, after one day, the error would be about twenty-four minutes of longitude; after ten days, this range would have increased to 1°16', and after a hundred days at sea, the accumulated margin of incertitude would have reached four longitudinal degrees. Finally, the modeling algorithm quantified these estimates of navigational error in terms of their effect on the confidence attributed to the associated magnetic observations.

The second major source of uncertainty stems from the process of observation; compass measurements were prone to many kinds of error. Some were caused by construction defects of the instrument, others by sighting impediments. Human fallibility also affected accuracy, during the taking of the reading, its recording in the log, and the modern conversion into machine-readable form. As earlier stated, magnetic declination was mainly observed at four instances during the nautical day: in the afternoon (azimuth), at sundown (amplitude), at sunrise the following morning (amplitude), and when the Sun again stood above the horizon (azimuth). At any of these times, multiple instruments or a quick succession of readings resulted in a series of up to about

fifteen measurements. The logbook frequently lists these quantities, but a calculated average figures only rarely. Navigators appear instead to have often resorted to taking the median of such a series.

By taking either individual readings or medians compiled on each day, their
spread around that day's mean can be quantified. As outlined before, a total of
18,918 observations recorded on 8,491 days yielded an overall standard deviation of 0.46 decimal degrees. The mariner was thus able to reduce the error
margin associated with individual measurements by taking advantage of multiple sighting opportunities in the course of a nautical day. The remaining uncertainty was directly factored into the modeling equations, joining the earlier-
defined crustal and positional error terms. Together, they represent the most
sophisticated effort yet to account for these diverse disturbances.

Building a Geomagnetic Model

A global reconstruction of the Earth's magnetism involves the application
of a number of mathematical techniques and physical principles, so as to describe the electromagnetic field (using Maxwell's laws) in a spherical, three-
dimensional framework. Only the core field was of interest here, so crustal sources
could be considered noise. This error factor effectively placed a lower bound on
the size of resolvable features at the Core-Mantle Boundary, thereby limiting the
number of model parameters to 224 at any single point in time. Furthermore, in
the temporal domain, 45 spline functions jointly bridged the period 1590–1800.
The full model consequently comprised a total of 224 times 45 equals 10,080
parameters, calculated with a technique known as *weighted damped least-
squares*. This required the manipulation of large datasets and considerable
computing time. The effect of any remaining improbable measurements was
limited by means of a clamping scheme, a method that gradually rejected outliers from the estimation process at successive modeling iterations.

Over a period of about eight months, several computers at the University of
Leeds produced a range of possible solutions. The complexity of the models
could be regulated by adjusting the level of *damping*. This constraint determined the amount of energy incorporated in the reconstructions. Certain limits confined them to a realistic range; very large damping would create almost
featureless results, whereas very small damping would allow the field to display
more detail than even modern satellite surveys are able to discern. A solution

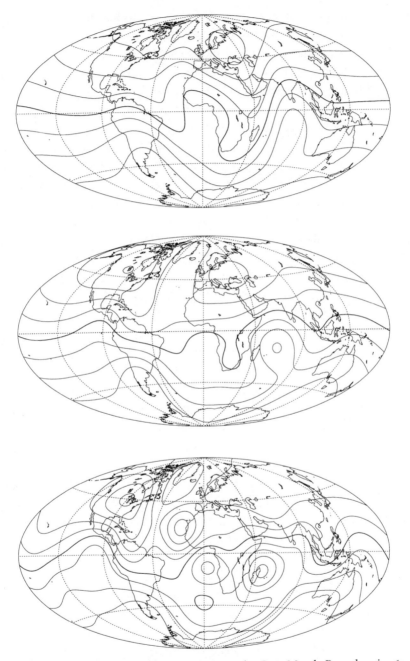

Fig. C.4. Modeled radial magnetic intensity at the Core-Mantle Boundary in 1600, 1700, and 1800, on a Mollweide projection. Contour interval = 12,500 nanoteslas.

closest to the "ideal" observational misfit of one standard deviation eventually became the basis for the interdisciplinary analysis.

The final outcome of the above-mentioned efforts is probably best appreciated in a three-dimensional animation. Here, it is presented as a series of snapshots of the field through time. Three global images of the radial component at the Core-Mantle Boundary display conditions for 1600, 1700, and 1800 (see fig. C.4), while three charts at century intervals from 1600 to 1800 map surface declination (see figs. 3.1, 4.1, and 4.4). These field reconstructions offer a significant improvement to previous efforts, both in time span and resolution of features. Moreover, the results appear generally consistent with previous models for the eighteenth century, derived in the past from much smaller datasets. Correspondence is, for instance, evident in the low intensities around the geographical poles, the four large "flux lobes" at high latitudes over Canada and Siberia, and their counterparts in the southern hemisphere. It will be recalled that this pattern is believed to be indicative of dynamo action in the outer core.

Many features are more sharply defined than in all earlier attempts to model the field over time. Among the static parts, the flux bundle in the Indian Ocean is, for instance, more concentrated than previously thought. Of the changing features, the movement of core spots can presently be analyzed in much greater detail than ever before. Especially the disturbance off the East African coast appears to have been imaged well; already visible in 1590, it gradually drifted west over decades toward Central Africa. In addition, the time dependency of the Canadian flux lobe is notable: this is not a static feature, but exhibits a wave-like motion which is most easily viewed in the form of an animation made from a series of images.

The three snapshots of the radial field also show that the degree of detail tends to increase with time. This is an artificial effect, due to the fact that earlier epochs are generally less well represented in the dataset. It is an unavoidable consequence of the paucity of the oldest sources, in combination with the fact that oceanic shipping itself grew over time from very modest beginnings. Nevertheless, even the reconstruction for 1600 shows remarkable detail, including an intensity high over Europe that appears to lose its strength only in the eighteenth century. The magnetic equator (the line separating outward and inward flux) seems, moreover, to be erratically undulating above and below the geographical equator, while also drifting westward. Field anomalies in the Atlantic and Indian Oceans are most clearly identified in 1800 as the possible cause.

The most interesting series of stills from the standpoint of the historical discipline are the isogonic charts. Such representations of declination depict the modeled global distribution of compass behavior at the surface, in contrast to the radial field component, which traced outward magnetic intensity at the Core-Mantle Boundary. Isogonics drawn at five-degree intervals designate the areas of northeasting and northwesting with which seventeenth- and eighteenth-century navigators, hydrographers, and natural philosophers became familiar. The movements of the isolines have been captured in steps of one hundred years, at each whole century between 1600 and 1800. Concentrating first on the two agonic lines (drawn in bold), around the turn of the seventeenth century, the Pacific one was grazing Japan, the Philippines, and New Guinea. In the next hundred years, its southern end slowly edged across Australia, while the northern part was making slow progress along the Chinese coast. Only in the second half of the eighteenth century did this agonic undergo a marked and rapid alteration, developing a large southwesterly wiggle through Borneo and the Sunda Strait toward the Indian subcontinent, where it continued roughly northward toward the pole.

The Atlantic agonic, on the other hand, seems to have been most unpredictable around the mid–seventeenth century. In 1600 it was still oriented almost meridionally across Africa, while in large parts of the Atlantic the needle pointed nearly true. Fifty years later, the agonic line had both drifted westward in the southern hemisphere and bent westward in northern quarters, via Spain and Ireland toward Nova Scotia and Hudson's Strait. By 1700 its shape had once again radically changed, then smoothly directing itself southwesterly from the mid-Atlantic toward and across North America. It more or less maintained this orientation during the next hundred years, while it continued to shift westward, reaching Brazil before the turn of the nineteenth century. As a result of these movements, the North Atlantic displayed very low declination in the seventeenth century, giving way to a largely parallel southwesterly pattern thereafter. This distribution can help to explain the sixteenth- and early-seventeenth-century confusion regarding the location of the agonic prime meridian, thought to run across (or near) the Azores, the Cape Verdes, and the Canary Isles. Given the poor instruments at the time, practically the whole region could yield a value of zero degrees difference between magnetic and true north.

In the South Atlantic, a system of curves centered on increasing declination in high latitudes slowly traveled in the direction of South America for over two hundred years. Particularly around that continent's southern and western edge did northeasting increase as the high moved further west. Finally, the isolines in the Indian Ocean were dominated by the earlier-mentioned core spot, which traveled southwestward near Madagascar. The accompanying area of northwesting was already well developed in 1600, and continued to grow in size and intensity while moving toward South Africa. At the turn of the nineteenth century, local declination east of the Cape had reached a maximum in excess of twenty-five degrees, whereas at destinations in the East Indies it had shrunk to near zero again (see above). The mariner was thus most challenged in this part of the world as regards his adaptability to local field characteristics, having to adjust his steering compensation almost on a daily basis.

Regrettably, some aspects of the model remain less than completely satisfactory. Where data are absent, information about particular features, even if clearly present at a later time, cannot simply be extrapolated into the past. Although it is possible to mathematically force a solution to be equally complex throughout the modeled interval, this would create spurious features where data are scarce, while muffling signal where and when observations abound. Reconstructions for the early seventeenth century will therefore always contain less detail than those for the late eighteenth century. Other weaknesses more directly associated with this family of solutions pertain to gaps in coverage. Even though declination readings yield information beyond the immediate vicinity of the observer, far too little was present to adequately model the Pacific. The Siberian coast and high southern latitudes equally suffered from a very low incidence of maritime traffic in the analyzed period. The reconstructions for these regions should therefore be considered potentially unreliable. The Siberian flux lobe appears rather poorly resolved, for example.

Fortunately, the very fact that so few of the processed logbooks described shipping in these remote regions implies that this omission is only of minor import for this study. The large majority of the material describes ventures in the Atlantic and Indian Oceans; these areas have never before been modeled so reliably, based on so many data points. In general, the new model constitutes an excellent representation of geomagnetic secular variation in the seventeenth and eighteenth centuries at the Earth's surface. Most of the available sixteenth- and seventeenth-century material has now been processed, and a full century (1590–1690) has been added to prior time-dependent reconstructions. Eighteenth-century coverage has similarly improved substantially. Furthermore, the proper quantification of the observational errors in the data, as well as the determination of the accuracy of navigation in the pre-chronometer era has not previously been investigated on the basis of a large sample of ships' logbooks.

In conclusion, the historical sources compiled in this study have significantly enhanced the geomagnetic reconstruction of the past. This clearly attests to the fecundity of the collaborative effort, at least from the geophysical perspective. Geomagnetists can now track secular variation for an unprecedented four hundred years (1590–1990), which allows them to study the slowly evolving patterns in greater detail than ever before. The findings may furthermore serve as a springboard to launch new hypotheses regarding the workings of the geodynamo in the present, the historical past, and on a geological time scale. Early-modern navigators have left modern geomagnetists with plenty of new horizons to explore.

The successful implementation of the presently chosen method, moreover, offers a firm foundation for the interdisciplinary analysis, to which the remainder of this chapter is devoted. Can clear distinctions be made between countries, periods, traversed areas, and maritime organizations, when comparing this physically constrained field model with historical estimates of local compass behavior? The answer will be based upon the paper legacy of hydrographers and navigators, who both dealt with the needle's variability in space and time. Their past labors have been laid down in sailing instructions and logbooks respectively, and it is to these sources that the present research has turned to compile two separate datasets.

Declination Predictions

A maritime-historical account of magnetic declination in sailing directions was featured in the previous chapter. The discussion is brief, and does not dwell on the attained precision of the predictions proper. Neither is such to be found in the existing scholarly works on the subject. There is a simple reason for this: the source material itself fails to carry this information. The scientific practice of accompanying estimates with error bars—an indication of the uncertainty margin—is quite recent, and sailing instructions of the time did not discuss this aspect at all. Navigation textbooks likewise remained completely silent on the topic, as did almost all navigational logbooks seen. Theoretically, one might have expected the latter category to yield substantial insight into this matter, but in practice even reference to a particular set of directions used is exceedingly rare, let alone discussion of the merits of individual remarks contained therein. The historian is thus left with a bare list of geographical locations and their associated northeasting or northwesting as predicted at a certain time, and there the traditional story ends.

A few examples may convey a flavor of the factual style of these documents. Around 1728, Knapton's *Atlas Maritimis et Commercialis* was quite economical with words when advising navigational practitioners on the port of St. Julian in Argentina: "The entrance into the harbour lies in lat. 49 d. 10 m, longit. from the Lizard 63 d 10 m . . . variation of the compass 16 d 10 m east."[3] French hydrographer d'Après de Mannevillette was more verbose, for instance in his 1765 printed *Mémoire sur la Navigation de France aux Indes:*

> Sailing toward the Cape of Good Hope, the variation, when one is able to observe it, is of great help in these waters to know approximately one's distance from this Cape. I have measured it in 1752 [to be] 19 degrees northwesting in the road of Table-Bay, and I believe it is presently more than 20 degrees . . . [I]t increases further [when sailing] toward the east up to 26 degrees; and following my observations and those that I have received since, this maximum variation is found near the longitude of the middle of the Mozambique Channel. It subsequently diminishes [when] going eastward.[4]

Lastly, a Dutch example from the East India Company guidelines issued in 1783 discussed the problem of establishing position relative to the islands of St. Paul and Amsterdam (in the Indian Ocean, on the route to Java): "Westerly and out of sight of these islands no less than 19 degrees is presently found, and near them usually 18 degrees of northwesting; but when the deflection has reliably decreased to 17 degrees, one can be assured to have passed the islands."[5]

Such published values, usually resulting from the examination of many recent logbooks, provided only a momentary fix, generally losing their edge within a few years, depending upon reigning secular acceleration at the reviewed sites. Frequent updates, published by maritime organizations during the second half of the eighteenth century, testify to a growing awareness that the battle against time could never be decisively won, and that keeping track of varying field changes required both a constant vigil and considerable expense. However, the question of how successful this struggle was has remained unanswered up to now.

The currently available geomagnetic model spans the period 1590–1800, and

Table C.1 The Accuracy of Sailing Directions, per Country and over Time

Period	English		French		Dutch		Total	
	N	s.d.	N	s.d.	N	s.d.	N	s.d.
1590–1650	49	12.722	0		101	5.258	150	8.723
1651–1700	21	6.815	94	4.984	67	7.947	182	6.657
1701–1750	47	6.490	34	4.072	63	3.964	144	4.936
1751–1800	23	2.449	67	2.476	138	3.058	228	2.845
Total	140	9.197	195	4.617	369	5.007	704	6.093

Note: N = sample size (the number of geomagnetic references gathered from various sailing directions); s.d. = standard deviation (their overall statistical spread around the modeled value for each stated time and place).

was specifically constructed as a quantitative tool with which to address this issue. It needs to be stressed that no spatial restrictions apply; regardless of whether a particular place has ever hosted observers of the Earth's magnetic field, the model can estimate all field components at any location above the Core-Mantle Boundary, and at any time within the modeled interval. Limitations in the temporal span of data covering large portions of the world regrettably forced the lower limit of 1590, thereby excluding earlier, mostly Iberian sources from the analysis. A sample of sailing directions by the three major maritime powers under scrutiny did, however, provide enough material to warrant confidence in the conclusions. A total of 22 English, 27 French, and 48 Dutch sources spawned a database of 704 references that satisfied the criteria of identifiable geographical location, clear date, and magnetic declination. Dedicated software converted these quantities to decimal degrees, made all longitudes relative to Greenwich, queried the model for the appropriate reconstructed declination value for each stated time and place, and computed their difference with the original historical estimate. For each source, the mean and standard deviation (measuring the scatter around the mean) of these residuals were subsequently calculated and tabulated per country, and divided among four epochs: 1590–1650, 1651–1700, 1701–1750, and 1751–1800.

As can be gleaned from table C.1, the overall sample is a modest one, and not every tabulation cell is equally well represented. Dutch material alone accounts for over half the total, whereas French sources in the first half of the seventeenth century, for example, are totally absent. Note also the differences within each country over time; the first epoch (1590–1650) is numerically strongest for the English, the second epoch for the French, and the fourth for the Dutch. Lastly, a stark contrast exists between the increased mention of declination in eighteenth-century French and Dutch sailing directions and an apparent parallel English decline.

For the most interesting and revealing figures, however, one has to turn to the residuals themselves. Starting with the total average, the standard deviation of all 704 residuals from the model amounts to about 6.1 (decimal) degrees. This average better reflects Dutch practice (reaching 5.0 degrees) than that of the English (at almost 9.2 degrees) and the French (4.6 degrees). The differences between epoch totals are even wider than those between nationalities, ranging between 8.7 and 2.8 degrees, with the first half of the seventeenth century the worst, and the second half of the eighteenth the best. Hydrographers clearly became better at this prediction game over time.

Closer inspection of the national standard deviations over time reinforces this notion, but also shows a difference between countries, with England reducing the averaged residuals more (10.3 degrees) than the French (2.5) and the Dutch (2.2 degrees). Most of this improvement in the English predictions took place from the first to the second half of the seventeenth century, while the French increased their accuracy most around a hundred years later. The Dutch sample, despite being the largest and boasting the best performance in the first epoch, nevertheless appears the least consistent, exhibiting considerable fluctuations over time. A clear trend is nonetheless visible in the convergence of levels of accuracy in all countries surveyed. The range between the three powers reduced from circa 7.5 degrees in the early seventeenth century to 0.6 degrees in the late eighteenth. Regarding absolute rankings, England and the Dutch Republic changed places three times, due to the Dutch decline in accuracy during the second half of the seventeenth century, and a slowly improving performance over the next ten decades. England and France eventually reached the highest precision throughout the eighteenth century, with excellent standard deviations of 2.4 and 2.5 degrees respectively. At that time, the Dutch were trailing at about 3.0 degrees. The French record is remarkable, given their late arrival on the international oceanic stage. But like no other, they apparently were able to assimilate existing knowledge quickly and improve upon it, leaving behind their Dutch competitors.[6] The distinction appears to mimic the one earlier described for observational accuracy, English and French readings being substantially more accurate than Dutch and Danish efforts. In summary, the three most important findings are:

1. Taken over the complete interval 1590–1800, sailing directions were not very accurate in describing the local field.
2. General precision markedly improved in the eighteenth century, notably during the latter half.
3. The French overall record is the best, followed by the Dutch; the English improved most, and eventually reached the highest accuracy.

The question remains why Dutch and French hydrographers did so much better than their English colleagues in predicting declination, and why the overall precision tripled over the period investigated. Both these points are directly related to the ability to collect and disperse magnetic data. The successful implementation of previously obtained knowledge on local declination and secular variation is primarily dependent on the uninhibited exchange of information. An efficient infrastructure thus needs to be in place to collect data in quantity, to distinguish signal from noise in stable and changing field features, and to distribute these findings among navigators.

Earlier it was mentioned how the logbook evolved from a personal diary of maritime matters into a highly formalized data carrier with preprinted sheets. Processing the information contained therein required a maritime organization to maintain some form of hydrographical office, as well as regulations inducing the handing in of logbooks soon after the completion of a voyage. Both the Dutch and the French implemented practical solutions; the former country through the VOC's official "chart maker" (from 1602) and several committees of examiners of masters, the latter by founding the naval "General Depot of Charts and Maps, Logbooks, and Accounts Concerning Navigation" in 1720. Forty-two years later the Compagnie des Indes established a similar office, headed by d'Apres de Mannevillette. Moreover, monetary sanctions upon failure to submit all relevant papers became effective by the eighteenth century in the navies and East India companies of both these countries. Similarly, the distribution of the extracted information was consciously driven by the perceived need to keep abreast of field change and amendments to earlier assumptions. The same motivation fueled the improvement of charts and the introduction of technical innovation in compass navigation.

The marked increase in precision of declination estimates in sailing directions during the second half of the eighteenth century can largely be attributed to this well-oiled machine, which was able to process hundreds of logs, and to reissue existing instructions with declination values adjusted accordingly. In this respect, it is instructive to recall table C.1 above, which shows that the growth in the number of geomagnetic references in such documents was purely due to developments in France and the Dutch Republic, while English notes on the field were dwindling. Most of the Dutch improvement in accuracy took place around the turn of the eighteenth century, whereas the French difference between epochs was greatest before and after 1750. In the former case, the examiners of masters in the Dutch East India Company may have played an important part, whereas in the latter, French officials such as d'Après de Mannevillette and Bellin have clearly been influential too. All worked in organizations that imposed strict controls to ensure the efficient processing of magnetic data.

If the infrastructure on the European mainland is compared to that on the other side of the Channel, it becomes obvious that conditions for hydrographical information gathering and dispersal by official bodies were far less favorable in England. It will be recalled that the English East India Company had briefly tried to copy its navigators' logbooks in the period 1614–28. But eventually it had abandoned issuing standardized sailing directions, giving the master merely his commissioned destination instead. Only in the second half of the eighteenth century did the Royal Navy design a separate form for mariners to use in storing maritime data (including local magnetic declination), which, however, was not officially introduced on board ship until 1804. English navigators meanwhile conducted their affairs without the benefit of an internal hydrographical office—in the East India Company until 1770, in the navy until 1795. The early failure of the company to appreciate the full capacity of logbooks as carriers of highly relevant information (including secular variation) has undoubtedly handicapped rapid, broad awareness of significant changes in the Earth's magnetic field through their publications. Left to their own devices, English navigators either kept personal records for reference (seventeenth century), or used translated Dutch and French material (eighteenth century). English reliance upon this foreign intelligence underlines both its necessity in ocean navigation and the failure of home-based maritime organizations to provide an adequate alternative. Without efficient means to collect, process, and redistribute information about the changing field, the few geomagnetic predictions that were committed to paper fell woefully short of requirements for much of the investigated period; only in the last half of the eighteenth century did English precision improve substantially, even surpassing all competitors in the maritime arena. By that time, the majority of hydrographers in all countries surveyed had finally pushed the limits of accuracy down to within the range of declination estimates made aboard ship.

Compensation at Sea

Allowance for local magnetic declination formed an integral part of oceanic navigational practice in the seventeenth and eighteenth centuries. Compass correction is evident in logbooks as a value incorporated in the dead-reckoning calculations. In the Dutch case, the rectifier offered the helmsman an additional physical means of aligning the north of the fly with the local meridian. This practice was often executed in steps of fixed size (for instance, 5 or 2½ degrees). Those navigators not dependent on this device retained more freedom to adjust for whatever estimate they deemed reasonable. But regardless of the manner of implementation, the question remains how these professionals arrived at a particular compensation at a specific place and time. The most obvious answer would be: from direct observation. Surely, this was the best policy to keep up with unpredictable changes, and there are indeed a few (late-eighteenth-century) cases where every opportunity to sight the Sun seems to have been exploited. But these are the exception rather than the rule. There are many logs that feature dozens of corrections and few or no readings at all, and even more voyages during which observational zeal was far from optimal. Those measurements that were taken often appear to have served merely to confirm an already established mental image of local geomagnetic conditions. The link between magnetic readings and compass adjustments was at times indeed a rather tentative one: only some 17 percent of all collected declination corrections coincides with a measurement obtained on the same day.[7] Even more poignant in this respect is the fact that a mere half of a percent actually corrected for the exact value observed. Clearly, the navigator had alternative sources at his disposal with which to estimate local declination.

As previously stressed, secular variation was fundamentally unpredictable over the time span of interest to the navigator. A mental image of altering field conditions would therefore require a sustained program of empirical checks to remain up-to-date. The information would also have had to be available shortly after procurement, since it would lose most of its value within a few years. This implies that several sources of geomagnetic data cannot have been very beneficial to the mariner in this respect. The first of these is the theoretical education of masters and mates. Some late-eighteenth-century navigation exams used by the Dutch East India Company did test memorized knowledge of locations and their associated northwesting or northeasting. In the best possible case, these figures would have been based upon the most recent sailing directions. But as the preceding discussion has already made clear, the time necessary for their production would tend to reduce their accuracy, and declinations learned by rote surely would not remain fresh for long. For the same reason, one can confidently exclude dependence on isogonic charts, and tables of local needle deflection as published in textbooks. Both are completely absent from lists of officially issued navigational tools, and not a single examined sailing direction makes any mention of them. The ability of the latter instructions to convey geomagnetic data has just been exposed as severely wanting in quality for much of the period under discussion.

These limitations in accuracy highlight the fact that only one potentially reliable source of information on the changing field existed at the time, and that was the Earth itself. A lifetime of shipboard observations constituted a far more viable means of tracking secular variation than readings published long after the fact. The important role of magnetic declination as a positional indicator underlines the mariner's efforts to consciously link certain needle behavior to specific locations. Such information need not necessarily have resulted from a frantic struggle to take advantage of every available opportunity to shoot the Sun. Although only a small proportion of compass corrections took place on a day that magnetic observations were made, there were, nevertheless, many alternative moments at which current knowledge could be renewed or extended: on the previous day, for instance, or at the same location on a previous voyage the year before. In addition to the current logbook, copies could be carried from journeys along the same traverse, either by the same person or by others. Private notes regarding declination are also found separately in maritime archives. These would all have functioned in tandem with the navigator's own memory, which was particularly trained to retain such information of geographical significance. Oral communication by colleagues, yet another means of transfer, was perhaps most practiced in a master-mate relationship on board, where the apprentice was being taught the ropes. A dockside brawl among seasoned officers over how much the needle northwested at the Cape somehow seems a less likely scenario.

Naturally, a mental representation of local declination along a ship's track could arise only as a result of repeated observations in known waters. Luckily, as far as routes were concerned, navigators received frequent instructions to travel along well-defined, narrow course corridors. Officers, moreover, often tended to remain employed on a given trajectory; English captains sailing to the East made more than four journeys on average, while many of their Danish and Dutch colleagues continued to ply the same ocean traverses for decades. Similar long records of service are also attested in some navy, merchant, and whaling logbooks.[8]

This leads to the intriguing question of how well seafarers accommodated for the field's peculiarities and inconstant changes, per country, over time, in different oceanic regions, and within maritime organizations. Fortunately, logbooks have the advantage over sailing directions in that they offer a stronger numerical basis for investigating their statistics. As such, a more robust sample offered itself for comparison with the geomagnetic model's best estimate for each recorded ship's position where a compensation was implemented. A total of 1,007 logbooks from England, France, the Dutch Republic, and Denmark eventually yielded 18,955 such compass corrections for analysis. The same software earlier used to study sailing directions then generated modeled declination values, residuals, and standard deviations (all in decimal degrees) for each triplet of ship's coordinates (latitude, longitude, and date) where a steering correction had been recorded. The far higher frequencies in this sample offered the chance to compare many subset categories; those that contained fewer than one hundred points have been disregarded. Tabulated results can be found in the appendix.

Starting with the total average, the overall standard deviation of all residuals from the model amounts to 2.46 (decimal) degrees. Navigational practice was thus on average more than twice as accurate as the sailing directions studied previously (at circa 6.1 degrees). But does this value have any bearing on historical reality? Does an average standard deviation of about two and a half degrees appear likely, from the standpoint of the early-modern northwest European mariner? Is it definitely too low, too high, or perhaps within reasonable expectation? Joseph Harris, "teacher of the mathematicks," in his 1730 *Treatise of Navigation* stated: "In correcting the courses it is sufficient that the variation be counted to a $\frac{1}{4}$ point."[9] A quarter point equals 2.8125 decimal degrees. Furthermore, a correction of two and a half degrees is ordered in numerous Dutch sailing directions from the latter half of the eighteenth century.[10] Other sources deemed half a point more feasible. Moreover, one standard deviation represents only about 68 percent of cases, whereas a range of two (or an accuracy of 4.9 degrees) covers more than 95 percent of the total. It therefore seems safe to

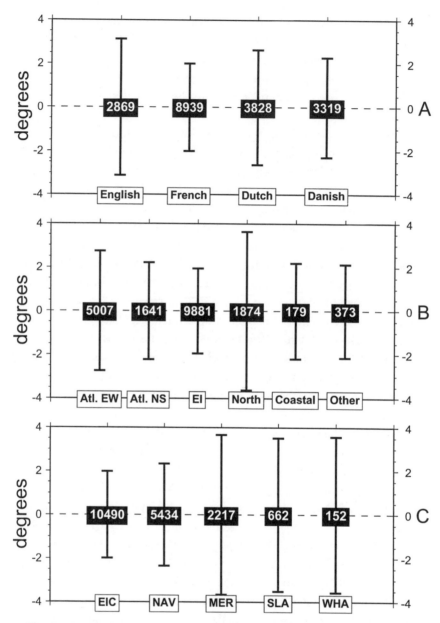

Fig. C.5. Standard deviations of residuals from the model for declination compensations at sea, in degrees. The line of zero degrees represents the modeled magnetic declination; the length of each vertical bar measures the range of residuals within one standard deviation from the model (ca. 68 percent); the boxed value on the zero line is the sample size. *A:* per country; *B:* per oceanic area; *C:* per type of maritime organization. For abbreviations, see the list in the front matter.

conclude that the computed levels of accuracy in allowing for local magnetic declination fall well within the plausible domain of navigation as practiced in the age of sail.

A closer inspection of subtotals reveals more detail than this preliminary result. The first examined division was made on the basis of nationality. Immediately apparent is the French domination, both numerically (accounting for nearly half the total) and in accomplished level of precision (2.0 degrees, against 2.3 degrees for the Danish, 2.6 for the Dutch, and 3.1 for the English competition). Thus, France and Denmark performed better than average, while the Dutch Republic and England disappointed in this respect.

Grouped by geographical range, East India voyages represent the majority of the material, with Atlantic east-west crossings coming second, and Arctic and north-south Atlantic traverses sharing third place. Coastal and other expeditions own only a modest share of the total. Remarkably, this last category ranks second (with 2.1 degrees, after East India runs with 1.9) in an ordered listing of precision, with data from Atlantic north-south traffic and coastal trips trailing around 2.2, journeys toward the Americas at over 2.7, and Arctic exploits at a poor 3.6 degrees.

The division into five types of maritime organizations was made before this analysis was performed. In advance, the distinction between navies and East India companies seemed obvious, as was the separation of whaling into an individual category. The assumption of slavers being different from the merchant marine, however, turned out to be more tentative. East India companies appear to have outperformed navies (at circa 2.0 and 2.3 degrees respectively), whereas the smaller sea-going enterprises followed at a considerable distance; the slave trade at 3.5 degrees, the whalers near 3.6, and the merchant marine coming last (3.7 degrees). In general, a large gap is thus in evidence between the more structured navies and East India companies, and the other maritime agents.

When investigated over time, the absolute number of corrections clearly shows an increase, with the largest growth taking place in the early eighteenth century. This effect is, however, caused to some extent by substantially more processed logbooks from those decades in the sample. During the seventeenth century, a slight improvement in accuracy is notable, from about 2.6 to 2.5 degrees. The best performance (less than 2.1 degrees) was evident in the first half of the eighteenth century, while during the latter half the standard deviation sagged back to over 2.6 degrees. The cause for this decline will be identified below.

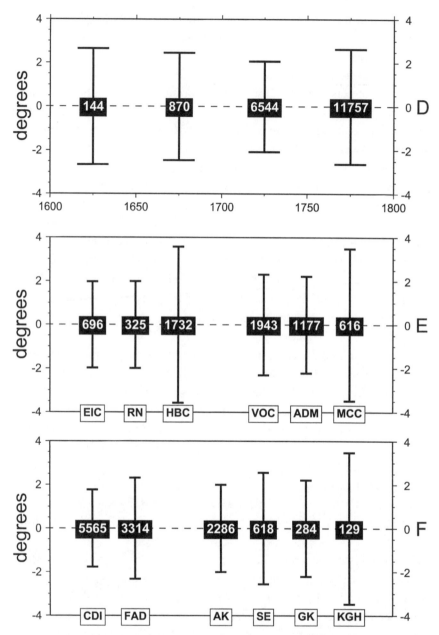

Fig. C.6. Standard deviations of residuals from the model for declination compensations at sea, in degrees. Zero line = modeled declination; vertical bar = one standard deviation; boxed value = sample size. *D:* over time; *E:* for English and Dutch maritime organizations; *F:* for French and Danish maritime organizations.

The last subsets to undergo scrutiny were the individual maritime organizations. Regrettably, several small companies failed to reach the lower limit of a hundred data points and had therefore to be excluded. Of those remaining, the French Compagnie des Indes and navy, the Danish Asia Company, the Dutch East India Company and navy, as well as the Hudson's Bay Company are most strongly represented in the sample, at over a thousand points each. Numerical predominance, of course, offers no guarantee of a top ranking in accuracy. It is possible to divide the population into three groups: those attaining a standard deviation at or below 2.0 degrees (containing the English Navy and all East India companies except the Dutch), those between 2.0 and 3.0 degrees (the French, Dutch, and Danish navies, the Dutch East India Company, and the Danish Guinea Company), and the unfortunate navigational practitioners unable to keep standard deviations below 3.0 degrees (of the English merchant marine and Hudson's Bay Company [HBC], the Danish Royal Greenland Trade Company, and the Dutch slavers [MCC]). Once again, the largest and most organized maritime institutions are seen to have performed best, with the Danish Guinea Company as exceptional addition. An interesting aspect is the superiority of East India companies over navies in all countries except the Dutch Republic.

Apart from analyzing results for each of the categories of subsets individually, it has also proved possible to follow developments within several combinations of them. They are:

country versus:
— epoch
— sailing route
— type of maritime organization
epoch versus:
— sailing route
— type of maritime organization

Other tabulations unfortunately led to a substantial overlap, and thus to a large number of nearly empty cells. This occurred because subdivisions of, for instance, organization and oceanic region are highly correlated; slavers did not travel to the Arctic, and ships of the Hudson's Bay Company never sailed the Indian Ocean. Following is a brief discussion of the findings for those combinations that did produce information. A more extensive tabulation of results is given in the appendix (see tables A.1–A.4).

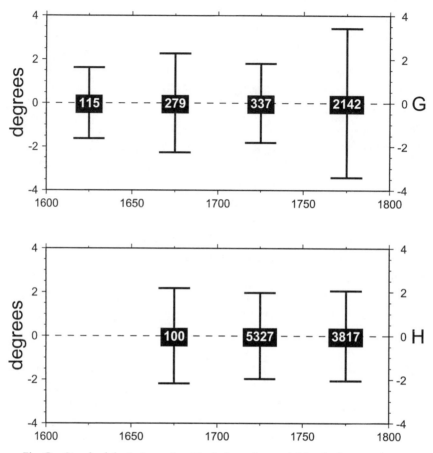

Fig. C.7. Standard deviations of residuals from the model for declination compensations at sea, in degrees, over time. Zero line = modeled declination; vertical bar = one standard deviation; boxed value = sample size. *G:* on English vessels; *H:* on French vessels.

To start with, French and Danish practice remained stable through time (the French at a very high precision, the Danes less so), whereas the English suffered a severe decline (from a very high initial standard) in the last half of the eighteenth century, due to the introduction in the sample of the Hudson's Bay Company. The Dutch, meanwhile, maintained a mediocre and erratic adjustment behavior throughout the two centuries surveyed. Regarding regional comparisons, large differences existed between English practice in the East Indies and elsewhere, and between Dutch Atlantic traverses in predominantly north-south and east-west directions, the former outdoing the latter in both

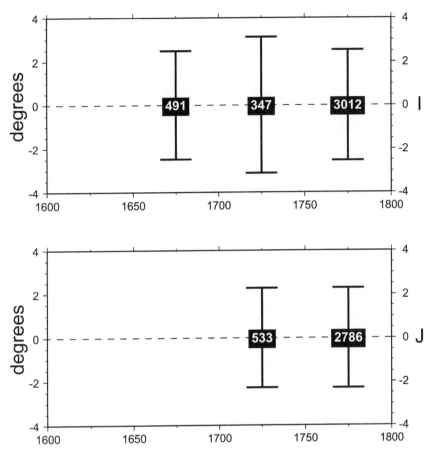

Fig. C.8. Standard deviations of residuals from the model for declination compensations at sea, in degrees, over time. Zero line = modeled declination; vertical bar = one standard deviation; boxed value = sample size. *I:* on Dutch ships; *J:* on Danish ships.

cases. Quite surprising are the small residuals on English meridional Atlantic traverses, and French coastal and exploratory journeys.

Among East India companies, the French also set the example, followed closely by England and Denmark, with the Dutch trailing somewhat behind. In a ranking of navies, the Royal Navy gained first place, followed by the Dutch and the French (about equal), and the Danes in last position. Merchant officers from Denmark appear to have had fewer problems with compass corrections than whalers from that country. Meanwhile, the accuracy maintained in different oceanic areas improved slowly with time, but without upsetting the initial order: East Indies, exploration destinations, Atlantic (north- and south-ward crossings outperforming east- and westbound voyages), and the Arctic closing the ranks. The variability of types of maritime organization over time was larger; both East India companies and navies improved their precision up to 1750, followed by a slight decline in the ensuing five decades.

As far as individual organizations are concerned, the English East India Company's sample was small, compared to that of the French Compagnie des Indes, but the attained accuracy in both was superb. Naval accuracy in both these countries displayed a little more fluctuation, and the French sample out-weighed that of the Royal Navy in size by over an order of magnitude. The Dutch East India Company long maintained a questionable level of expertise in tracking local field characteristics. Their eventual improvement in the sec-ond half of the eighteenth century, to a precision nearly equaled by the Dutch Admiralties, mirrored a slight increase in accuracy by the Danish Asia Com-pany during that time. Reliable data from the Danish Navy and many of the smaller companies are available only for the last fifty years of the studied pe-riod. No organization featured particularly small errors; in particular the slave trade and various merchant companies consistently displayed a quite lamenta-ble level of expertise in adjusting for local needle deflection.

Presently, the geomagnetic estimates of hydrographers and navigators have been statistically quantified for the first time. As a result, a puzzling phenome-non has emerged, namely the discrepancy in standard deviations between sail-ing directions and compass corrections at sea. Both professions depended on the same basic input of compass readings, and positional, observational, and crustal error plagued interpretation of the data for each. Moreover, hydrogra-phers and other experts on shore had the advantage of a far greater number of logbooks for analysis than the average captain or master had access to. One would therefore expect the former group to be much better able to winnow the grain from the chaff, by reducing individual error and identifying outliers. Yet the overall standard deviation of declination estimates in sailing directions was far larger than even those maintained in Arctic whaling. Was a hidden influ-ence perhaps not, or insufficiently, taken into account? Or did an inherent flaw exist in the hydrographical evaluation of geomagnetic information? Some-where along the line between experience gained and disseminated, something was lost in the translation.

The elusive enemy was, of course, time. Provided with adequate instru-ments, the navigator at sea could apply empirical data almost immediately

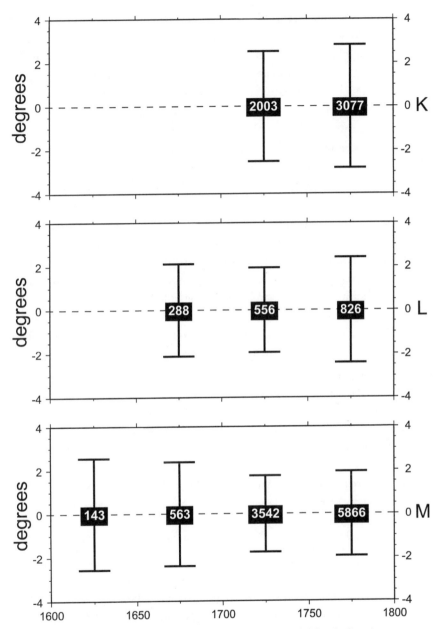

Fig. C.9. Standard deviations of residuals from the model for declination compensations at sea, in degrees. Zero line = modeled declination; vertical bar = one standard deviation; boxed value = sample size. *K:* Atlantic parallel courses (E-W); *L:* Atlantic meridional courses (N-S); *M:* to and from the East Indies.

after he had obtained them. If sailing regular routes, he could furthermore constantly update existing knowledge, swiftly incorporating noted alterations. The hydrographer, on the other hand, had to wait for the data to reach the home country, which could take years. Then he had to extract various kinds of potentially valuable information from the logs, of which geomagnetic readings formed only one, modest aspect. The reliability of each data point had to be established by comparison with others made nearby (in a spatial and a temporal sense), and a weighted estimate made for each location of interest. In the case of the French *Dépôt des Cartes et Plans*, a single individual probably performed this task, which did not expedite matters, on account of the large number of documents to be processed. The Dutch East India Company, conversely, arrived at conclusions regarding local declination through the extensive deliberations of a committee of experts, which equally failed to ensure rapid progress, because of differences of opinion. After final approval, the organizations had then to print and publish the text. Officers out at sea at the time might have had to wait another year or two before being able to lay eyes on the new instructions, depending on their travel itinerary. Meanwhile, all the time since the original measurements had been obtained, liquid iron continued to flow deep inside the planet, and magnetic features at the surface followed suit, changing shape, position, and intensity. These land-based efforts at field description thus became self-defeating prophecies, due to the expenditure in time necessary to draft and issue a new compilation. The bonus of error reduction through careful synthesis of numerous observations was thus more than outweighed by the benefit of direct experience on regular routes. Extrapolation of existing trends held no guarantee for the future; sailing directions could therefore report only what had once been observed. Chained to the past through relay by proxy, geomagnetic references in sailing directions would always remain behind the times.

Past Attractions

This book encapsulates the results of an investigation into the rewards of an interdisciplinary research strategy. Geomagnetism and history forged a temporary alliance to uncover new information of benefit to both. The goals to be achieved by each participant were well defined in advance, and needed little adjustment along the way. Nevertheless, some challenging obstacles had to be overcome to gain a measure of success. Both disciplines have learned from each other in the course of these investigations: geophysics gained insight into locating and processing maritime-historical sources, while history acquired a rudimentary knowledge of the power of global modeling. This cross-over is perhaps taken for granted, whereas each participant's methodology was in many ways quite alien to the other. They share a devotion to reconstructing the past through indirect means, but there the superficial similarity ends. Yet in spite of profound differences, enough mutual ground turned out to exist for practical, meaningful, and lasting collaboration.

The enterprise did involve a considerable expenditure in time, energy, and resources. Prolonged stays in repositories in England, the Netherlands, and France were necessary, in addition to shorter forays at many other sites. Bulk processing of sources is probably the only effective method to collect substantial datasets of historical observations of, and corrections for, magnetic declination at sea. The main problem here was psychological in nature. Correcting the compiled data had to await the completion of the archival retrieval phase, and the comparison of the field reconstruction with the historical estimates could be initiated only after the choice of modeling parameters had been finalized. In practice, this meant having to work in the dark for three years.

But eventually, the combined efforts generated the first time-dependent model of geomagnetic secular variation spanning four centuries (1590–1990), as well as the smaller model used in this particular historical analysis (1590–1800). Both were based on what has become, at the time of writing, the world's biggest compilation of historical magnetic observations. Substantially larger amounts of data and more sophisticated error estimates have allowed field features to be resolved both in closer detail and further into the past than ever before. The statistical analyses of error in navigation and measurement have, moreover, proved to be relevant not just in constructing an appropriate weighting matrix in the modeling process; the findings also benefited maritime history, which had never before undertaken such a quantified approach to the subject.

An unavoidable aspect of dealing with historical sources is the frustrating confrontation with their incompleteness. The majority of potentially useful material has not survived the ages, and the remainder is scattered very unevenly through time. In the present survey of logbooks, for example, nearly a thousand manuscripts represent the second half of the eighteenth century alone (sampled from an even larger array), whereas a thorough search of many archives and libraries has yielded only 142 such documents dated prior to 1651. Sadly, this lack of information for early times has meant that geomagnetic models tend to become less detailed the further back in time one reconstructs. This decrease in complexity is thus primarily related to poorer data coverage in these periods, rather than representing any real physical change at depth. Fortunately, this effect is largely confined to the Core-Mantle Boundary; a reconstruction of field patterns at the surface is mostly insensitive to small changes at their origin deep below. It is thus mainly a concern from a geophysical standpoint, whereas it actually served to make the historical studies more robust. This was also apparent in the available choice of solutions; a wide range of models all resulted in fairly uniform standard deviations in the comparative analysis of sailing directions and compass corrections.

Regarding the purely historical part of the research, the international perspective on geomagnetic navigation is perhaps the most important contribution to maritime studies. Parallel and disparate improvements have clearly influenced neighboring countries, while at the same time maintaining an individual style shaped by reigning local conditions. Both the common and the peculiar were thus highlighted in colors beyond the spectrum of more geographically confined approaches. The compass has occupied the minds of many, and the mariner's reliance on the instrument involved much more than steering alone. Technological advances made it possible to measure the angle between true and magnetic north with remarkable accuracy, offering an intriguing and sometimes confounding complication in getting from departure to destination across the oceans. But the mariner was quick to turn this handicap into an asset, by associating particular values with specific places for certain times. Sailing directions, navigation manuals, tables, and even some charts came to include references to the changing field, in addition to thousands of logbooks consulted by hydrographers and navigators alike. Magnetic declination was probably a blessing in disguise, requiring substantial extra efforts in observation and compensation, but also rendering a little reassurance in a sea of unknowns.

The advantage of the international approach is even more clearly apparent in the survey of the evolution of geomagnetic hypotheses. The various solutions to the longitude problem involving terrestrial magnetism, as well as the theories propounded to explain the observed phenomena, were almost as diverse as the background of their inventors. A chronological arrangement of many dozens of such postulates has enabled the identification of, and attribution to, four distinct phases of complexity. The definition of these phases then allowed the separation of defined trends from idiosyncrasies in geomagnetic thought for the very first time.

Moreover, fundamental ideas regarding cosmology, the inner structure of the planet, corpuscular, effluvial, and ethereal magnetic fields, as well as the detection, representation, and prediction of patterns on a global scale lie at the root of this inceptionary phase of the geomagnetic discipline. It still seems surprising in hindsight that so many of these thoughts were motivated by navigational concerns, and built upon observations made at sea by practical professionals, most of whom were themselves not at all concerned with the theoretical implications of deep-Earth conjectures. Even more remarkable is the fact that the communication between these different professional spheres was in some cases bilateral; some special compasses and magnetic longitude schemes were tested at sea, and in a few cases even temporarily adopted. Even though all participants were pursuing their own agendas, there was a considerable area of overlap and interaction.

The third group of professionals that attempted to make sense of the field's intricate characteristics was made up of hydrographers. The predictions of local declination contained in their sailing directions formed the first, modest part of the main purpose of the project: the comparison of modeled declination with historical geomagnetic estimates. A more extensive sample of compass compensations was extracted from practical navigation, fleshing out the maritime record in several arrangements of subsets. The analysis of their standard deviations were quite revealing: sharp contrasts have come to light between countries, periods, traversed areas, and maritime organizations. Some of these were not surprising: navigators often had a firmer grip on the straying needle than hydrographers, East India companies and navies tended to perform on a higher level than smaller merchant ventures, and accuracy generally increased over time. On the other hand, England's poor showing in sailing directions up to 1750, France's excellent performance across the board, the difference between east-west and north-south Atlantic traverses, and the variable record of the Dutch East India Company were all quite unexpected.

These varied aspects of humankind's historical struggle to come to terms with a fundamentally unpredictable phenomenon could only have been illuminated to the present extent by engaging in an extensive dialogue with the geophysical discipline. Without a global geomagnetic model, none of the residuals could have been calculated, and no differences could have been statistically quantified. In hindsight, the findings thus clearly support the fundamental premise: the synthesis of modern geomagnetic and historical field descriptions does indeed bear fruits. Clear differences in accuracy confirm that the employed model of quantified change can be used profitably to examine the historical past, in ways reserved to those who invest in an interdisciplinary research strategy. The answer to the fundamental question posed at the end of this chapter's introduction thus has to be affirmative.

In a more general sense, the investigation furthermore emphasizes the necessity of taking sizeable samples from the available sources, in order to separate signal from noise. A hidden layer of information may thereby be uncovered, which awaits recognition beneath the surface of individual documents. When linked with additional expertise from other disciplines, scattered data may then turn out to form clusters, and even error and inaccuracy may hold information about the past. This method is not meant to replace or belittle more traditional approaches to studying history, but merely to supplement the existing tools of analysis. Numerical modeling of time-dependent, quantifiable phenomena is certainly not restricted to geophysics, and may very well grant access to many other areas previously thought to be out of bounds for historical research.

Appendix

The four tables in this appendix list frequencies and indicators of attained accuracy in geomagnetic navigational practice, for a number of defined subsets. The data have been analyzed per country, period, route, and type of maritime organization (table A.1); the main types of organization, per country (table A.2); period and traversed oceanic region, per country (table A.3); and route and type of organization, over time (table A.4). The first two tables columnize: the number of processed logbooks, the number of extracted geomagnetic observations used in the time-dependent model, the compensation step size (see below), and the recorded allowances made for local magnetic declination in steering, stating the number of data (N), and the standard deviation (s.d.) of their residuals from the geomagnetic model. The last two tables columnize for each subset concerned: the number of steering allowances (N), and the standard deviation (s.d.) of their residuals from the geomagnetic model. All angular values are expressed in decimal degrees. For a discussion of compass allowance, see chapter 6; the modeling and statistics are treated in the concluding chapter; some results have been visualized in figures C.5–C.9.

The compass "step size" is defined as the angular difference between two successive steering corrections in compensating for local magnetic declination at sea (briefly discussed in chapter 6). For each logbook containing two or more of such declination corrections, the minimum step size between successive compensations was recorded. No maximum range in time was applied during a voyage, so some of the larger values may be unrealistically high; actual practice may well have been more accurate. Tables A.1 and A.2 columnize under "step size": the sampled number of logbooks that contained two or more compass allowances, and the average of all obtained minimum step sizes (one for each logbook).

Summarizing the main results, the overall average of compass correction step size amounts to 2.312 degrees. In terms of nationalities, the highest accuracy was reached by the French, followed by the Danes, the English, and (at great distance) the Dutch. The latter's poor result is, however, unfavorably affected by a significant proportion of seventeenth-century material. When viewed over time, a general increase in accuracy is counteracted only to some extent in the second half of the eighteenth century, when the sample starts to contain a far larger number of merchant logbooks. Especially the abundant data from a Dutch slaving company (the MCC; see table A.2) has had a detrimental effect.

In terms of traversed oceanic region, navigators bound for the North Atlantic and the East Indies tended, on average, to compensate for changes in local declination in smaller steps than on Atlantic voyages along parallels in moderate latitudes; in particular the poor performance on north-south Atlantic runs deserves attention. The good record for northern waters is largely maintained by navigators from the Hudson's Bay Company. When individual countries are compared, English and Danish material show large differences between east-west (parallel) and north-south (meridional) Atlantic crossings.

A comparison of types of maritime organization yields the surprising result that both navies and merchants (other than slavers) outperformed the East India companies; slavers and whalers follow at a considerable distance. Examined per country, the disparity between East India companies and navies is reflected in all individual countries except France; both the Compagnie des Indes and the French Navy exhibit excellent, almost identical records. The contrast is most pronounced in Denmark (at high precision) and the Dutch Republic (low precision). As indicated above, the poor performance of the slavers is caused almost entirely by the Dutch MCC.

Naturally, these statistics represent but one aspect of geomagnetic practice, and should be considered jointly with both the instrumental accuracy in obtaining local declination, and the tabulated standard deviations of the residuals (obtained by comparing the geomagnetic model at each specified time and place with the local compass compensation). To give but one example, in the very first division by country in table A.1, the commendable French practice of small (average) step size (about 0.7 degrees) correlates well with the lowest standard deviations of the allowances made (2.0 degrees), but no such straightforward relationship holds for the other countries. The Dutch, for instance, managed to maintain an average accuracy in their estimates of local declination, despite having the worst record of all in shifting their cards (see chapter 6). This shows that correction step size was not the only factor in the equation; their time and place may have been of even greater importance.

Table A.1 Statistics of Geomagnetic Navigational Practice:
Compass Observations and Steering Compensations (1)

Criterion	Subset	logs	obs.	logs	avg. min.	N	s.d.
				Step Size		**Allowance**	
Country:	England	601	11,567	139	2.079	2,869	3.122
	France	567	17,313	298	0.680	8,939	2.000
	Dutch Republic	811	18,487	380	3.824	3,828	2.627
	Denmark	77	3,939	58	1.344	3,319	2.287
Period:	1590–1650	142	1,793	19	4.093	144	2.647
	1651–1700	290	5,700	114	3.844	870	2.453
	1701–1750	645	15,198	282	1.862	6,544	2.071
	1751–1800	985	28,708	460	2.134	11,757	2.633
Route:	Atlantic E-W	817	6,589	313	2.339	5,007	2.735
	Atlantic N-S	201	5,627	137	3.329	1,641	2.222
	East India	706	37,562	345	1.978	9,881	1.947
	Arctic	110	43	52	1.679	1,874	3.624
	Coastal	192	288	18	2.748	179	2.192
	Other	36	1,290	10	1.567	373	2.139
Type of	East India C.	796	33,676	423	2.185	10,490	1.984
Organization:	Navy	889	16,441	266	1.630	5,434	2.341
	Merchants	151	400	73	1.789	2,217	3.662
	Slavers	156	737	105	4.812	662	3.514
	Whalers	57	41	8	3.669	152	3.565

Note: E-W and N-S denote the general orientation of Atlantic traverses (parallel and meridional);
East Indian route encompasses the Atlantic and Indian Oceans; logs = total number of processed logbooks; obs. = number of observations; N = sample size; avg. min. = averaged minimum step size per
logbook; s.d. = standard deviation; C. = Company.

Table A.2 *Statistics of Geomagnetic Navigational Practice: Compass Observations and Steering Compensations (2)*

Country	Organization	logs	obs.	Step size		Allowance	
				logs	avg. min.	N	s.d.
England:	East India C.	137	6,122	58	2.769	696	1.976
	Royal Navy	390	5,326	28	2.271	325	1.989
	Hudson's Bay C.	45	20	44	1.210	1,732	3.569
France:	East India C.	248	11,742	165	0.682	5,565	1.768
	Navy	305	5,366	129	0.681	3,314	2.321
Dutch	East India C.	374	13,009	169	3.683	1,943	2.312
Republic:	Admiralties	175	4,715	93	2.726	1,177	2.219
	Slavers (MCC)	143	703	101	4.794	616	3.484
Denmark:	East India C.	37	2,813	31	0.925	2,286	2.000
	Navy	20	1,036	16	1.784	618	2.556
	Guinea C.	11	88	7	1.979	284	2.210
	Whaling (KGH)	4	0	3	1.355	129	3.478

Note: E-W and N-S denote the general orientation of Atlantic traverses (parallel and meridional); logs = total number of processed logbooks; obs. = number of observations; N = sample size; avg. min. = averaged minimum step size per logbook; s.d. = standard deviation; C. = Company; MCC = Middelburgsche Commercie Compagnie; KGH = Kongelige Grønlandse Handelskompagni.

Table A.3 *Statistics of Geomagnetic Navigational Practice: Steering Compensations per Country*

Subset	English		French		Dutch		Danish	
	N	s.d.	N	s.d.	N	s.d.	N	s.d.
1590–1650	115	1.625	0		29	4.729	0	
1651–1700	279	2.266	100	2.178	491	2.492	0	
1701–1750	337	1.810	5,327	1.963	347	3.116	533	2.279
1751–1800	2,142	3.410	3,817	2.052	3,012	2.529	2,786	2.281
Atlantic E-W	168	3.004	3,234	2.394	1,137	3.057	468	2.931
Atlantic N-S	92	1.636	664	2.125	679	2.414	206	1.800
East India	862	1.996	4,586	1.642	1,931	2.246	2,502	1.974
Arctic	1,732	3.569	0		18	3.186	124	3.953
Coastal	9	0.808	108	1.161	43	3.609	19	1.297
Other	6	3.182	347	1.981	20	3.426	0	

Note: E-W and N-S denote the general orientation of Atlantic traverses (parallel and meridional); East Indian route encompasses Atlantic and Indian Oceans; N = sample size; s.d. = standard deviation; C. = Company.

Table A.4 Statistics of Geomagnetic Navigational Practice: Steering Compensations over Time

Subset	1590–1650		1651–1700		1701–1750		1751–1800	
	N	s.d.	N	s.d.	N	s.d.	N	s.d.
Atlantic E-W	0		10	4.995	2,003	2.525	3,077	2.818
Atlantic N-S	0		288	2.119	556	1.938	826	2.408
East India	143	2.556	563	2.380	3,542	1.753	5,866	1.941
Arctic	0		0		34	3.900	1,840	3.610
Coastal	0		9	1.172	95	1.288	78	2.793
Other	1		0		314	1.923	70	2.948
East India C.	144	2.647	826	2.426	4,935	1.910	4,887	1.922
Navy	0		20	3.952	1,437	2.253	4,027	2.319
Merchants	0		24	1.511	75	2.314	2,122	3.713
Slavers	0		0		97	3.946	569	3.380
Whalers	0		0		0		152	3.565

Note: E-W and N-S denote the general orientation of Atlantic traverses (parallel and meridional); East Indian route encompasses Atlantic and Indian Oceans; N = sample size; s.d. = standard deviation; C. = Company.

Chronology of Geomagnetic Hypotheses (1500–1800)

The table below lists, in chronological order, the geomagnetic hypotheses put forward between 1500 and 1800, as compiled from English, French, Dutch, and Spanish sources. Concepts originating elsewhere have been included when encountered, although no active search was undertaken to find them. Other schemes may therefore still await future (re)discovery.

The year of first publication (either in manuscript or in print) is followed by the proponent(s) and their region of origin. The description of the hypothesis consists of the general type (see chapter 2) and all unambiguous parameters rendered. Specific practical navigational aids, such as tables and charts, are mentioned separately where appropriate. The "Result" column designates whether a scheme was meant to find physical distance traveled along a parallel of latitude, the location of a magnetic mountain, longitudinal or latitudinal degrees, or was merely intended to explain or describe geomagnetism. All schemes completely lacking definite particulars have been omitted.

Year	Proponent(s)	Hypothesis	Result
1508	João de Lisboa and Pedro Anes (Portugal)	tilted dipole 180° E of Azores, declination-distance table	distance
1519	Francisco and Ruy Faleiro (Portugal)	tilted dipole 180° E of Azores	longitude
——	Filipe Guillen (Spain)	tilted dipole	longitude
1522	Sebastian Cabot (Italy)	celestial(?) tilted dipole colatitude 24°, 155° E of Canaries	longitude
1525	Antonio Pigafetta (Italy)	celestial axial dipole, tilted dipole	distance
1542	Alonso de Santa Cruz (Spain)	tilted dipole meridional isogonic chart	longitude
——	Jean Rotz (France)	tilted dipole	distance

1545	Martin Cortés (Spain)	celestial tilted dipole 180° E of Azores	longitude
1547	Gonzalo Fernandez de Oviedo (Spain)	tilted dipole 180° E of Azores	longitude
——	Gerard Mercator (Flanders)	tilted dipole colat. 11°, 168° E of Canaries	magnetic mountain
1550	Hieronymus Cardano (Italy)	celestial tilted dipole	
1558	Giambattista della Porta (Italy)	tilted dipole	longitude
1562	Jean Taisnier (France)	celestial axial dipole	
1569	Gerard Mercator (Flanders)	tilted dipole colat. 13°, 172° E of Azores; or colat. 16°30´, 178° E of Cape Verdes	magnetic mountain
1573	Pe(d)ro Menéndez de Avilés (Spain)	tilted dipole 180° E of Azores	longitude
1574	William Bourne (England)	tilted dipole	longitude
——	Toussaints de Bessard (France)	celestial tilted dipole declination-distance table	distance
1581	William Borough (England)	tilted dipole colat. 25°44´, 180° E of Azores	longitude
——	Robert Norman (England)	nuclear axial dipole	latitude
——	Michiel Coignet (Flanders)	tilted dipole colat. 16°30´	lodestone mine
1583	Jacques de Vaulx (France)	tilted dipole 179° E of Canaries	distance
1590	José de Acosta (Spain)	quadrupole	longitude
1595	A. Helmreich von Eissfeldt (Germany)	tilted dipole	longitude
1596	Francisco da Costa (Portugal)	quadrupole meridional agonics through the Azores, Cape Agulhas, Pedra Branca, Canton, and Cartagena	longitude

1597	Thomas Blundeville (England)	tilted dipole colat. 16°22′ or 25°44′, 180° E of Azores	longitude
1598	Petrus Plancius (Dutch Republic)	quadrupole colat. 23°30′, meridional agonics at 0°, 60°, 160°, and 260° E of Azores	longitude
1599	Simon Stevin (Dutch Republic)	sextupole colat. 23°30′, agonic great circles at 0°, 60°, and 160° E of Azores	
——	Edward Wright and Henry Briggs (England)	axial dipole dip-latitude table	latitude
1600	William Gilbert (England)	axial dipole, crustal irregularities dip-latitude diagram	latitude
1602	Pedro de Syria (Spain)	tilted dipole colat. 4–5°, 180° E of Azores	
——	Thomas Blundeville (England)	axial dipole dip-latitude table (Briggs 1599)	latitude
1603	Guillaume de Nautonier (France)	tilted dipole colat. 23°, 180° E of Canaries	longitude, latitude
1607	Bartolomeo Crescentio (Italy)	tilted dipole 180° E of Azores	longitude
1608	Luis de Fonseca (Portugal)	tilted dipole 180° E of Azores	distance
1609	Anthony Linton (England)	tilted dipole	longitude
——	Barent Evertsz Keteltas (Dutch Republic)	quadrupole, axial dipole (Plancius 1598; Briggs 1599)	longitude, latitude
1613	Mark Ridley (England)	axial dipole (Briggs 1599; plus his own table)	latitude
1615	Lorenzo Ferrer Maldonado (Spain)	tilted dipole	longitude
1621	Jean Tarde (France)	tilted dipole colat. 23°30′	longitude
1624	Valentim de Saa (Portugal)	quadrupole meridional agonics near the Azores, Cape Agulhas, Pedra Branca, and Vilalobos	longitude

1625	Manuel de Figueiredo (Portugal)	quadrupole meridional agonics near the Azores, Cape Agulhas, Canton, and Vilalobos	longitude
1620s	Cristovao Bruno (Italy)	quadrupole meridional agonics near the Azores, Cape Agulhas, Pedra Branca, and Acapulco (first manuscript chart with curved isogonics, lost)	longitude
1634	Henry Gellibrand (England)	discovery of secular variation at London (published 1635)	
1635	Nathaniel Carpenter (England)	axial dipole, crustal irregularities dip-latitude table (Briggs 1599)	latitude
1639	Henry Bond (England)	atmospheric precessing dipole colat. 8°30′, at longitude of London in 1657, clockwise orbit, 600 years	longitude
1640	Nicolas Le Bon (France)	celestial quadrupole colat. 23°30′, agonic meridians near the Azores, Cape Agulhas, Pedra Branca, and Vilalobos	longitude
1641	Athanasius Kircher (Germany)	celestial axial dipole, magnetic fibers, dip-latitude table	latitude
——	anonymous (England)	tilted dipole colat. 11°15′, longit. of Moscow	longitude?
1645	Jacques Grandamy (France)	axial dipole, crustal irregularities dip-latitude table	latitude, longitude
1647	Gabriel Grisly van Offenburgh (Dutch Republic?)	quadrupole	longitude
1652	Johannes (Phocylides) Holwarda (Dutch Republic)	axial dipole dip-latitude (Kircher 1641)	latitude
1655	Antonio de Mariz Carneiro (Portugal)	quadrupole	longitude

1659	Henry Phillippes (England)	precessing dipole longit. of London in 1657, clockwise orbit, 370 years	longitude
1666?	anonymous (England)	precessing sextupole (Stevin 1599) clockwise orbit with epicycles, 380 years	longitude
1672	John Seller (England)	axial dipole dip-latitude table (Briggs 1599)	latitude
1673	Luiz Serrão Pimentel (Portugal)	quadrupole (Mariz Carneiro 1655)	longitude
1674	Robert Hooke (England)	precessing dipole colat. 10°, "Pacific" longit., clockwise orbit, 370 years	
1680	Peter Perkins (England)	quadrupole/sextupole(?) orbit slower than in Bond 1639	
1683	Edmond Halley (England)	disjointed quadrupole see figure 3.3 and table 3.2	
1692	Edmond Halley (England)	crustal and nuclear disjointed dipole nucleus orbits clockwise relative to crust, 700 years	
1695	Edmond Halley (England)	manuscript isogonic chart of the Antarctic, lost	longitude?
1696	Edward Harrison (England)	crustal and nuclear precessing dipole nuclear pole near London longit., period 370 years; crustal pole on American longit., period 700 years	longitude
1701	Edmond Halley (England)	printed isogonic chart of the Atlantic Ocean	
1702	Edmond Halley (England)	printed world map with isogonics covering the Atlantic and Indian Ocean	
1706	Edward Howard (England)	tilted dipole	longitude

1710	Noel Feuillée (France)	double precessing dipole	longitude?
1718	Christoph Eberhard (Germany)	precessing dipole	longitude
1721	William Whiston (England)	nuclear disjointed dipole N-pole: colat. 13°30´, 30° E of London in 1720; ring-shaped S-pole: colat. 30°, 117° E of London; clockwise orbit, 1,920 years	longitude
1729	Zachariah Williams (Wales)	precessing dipole colat. 9°, 185° E of London in 1660; colat. 15°, 307° E in 1860, elliptical counterclockwise orbit, 591 years	longitude
1731	Jean-Philippe de la Croix (France)	precessing dipole	longitude
1732	Guillaume Le Vasseur de Dieppe (France)	tilted dipole colat. 23°30´, 180° E (of Paris?)	longitude
——	Servington Savery (England)	precessing dipole, core topography colat. 8°20´, longit. of London in 1657, orbit affected by Moon, 580 years	longitude
——	anonymous Jesuits at Ingolstadt University (Germany)	precessing dipole colat. 8°02´, 96°14´ E of Canaries in 1600, clockwise orbit, 300 years	longitude
1734	Arthur Bedford (England)	precessing dipole colat. 16°, longit. of London in 1705; clockwise orbit, 180 years	longitude
——	Emanuel Swedenborg (Sweden)	disjointed dipole colat. 22°30´, longit. from London in 1720 112° W (N-pole) and 145°30´ W (S-pole), counter-clockwise orbit, N-pole: 386 years, S-pole: 1,080 years	

——	Jeremy Woodyer (Ireland?)	tilted dipole colat. 30°, in the meridian of London	longitude
1738	Nicolaas Samuelsz Cruquius (Dutch Republic)	tilted dipole 2°30′ E of Canaries	longitude
1747	Feretti (France)	tilted dipole	longitude
1748	Gowin Knight (England)	tilted dipole colat. 66°30′	
1752	Mandillo (Italy)	double dipole axial and equatorial, the latter at 90° E and W of the longit. of Cape Agulhas	longitude, latitude
1753	Arnold Maasdorp (Dutch Republic)	precessing quadrupole	longitude
1755	Meindert Semeyns (Dutch Republic)	triple disjointed dipole colat. and longit. from Canaries in 1580: 28°36′, 268° E (N-pole, crust), 28°36′, 164° E (S-pole, crust), 19°30′, 136°40′ E (N, shell), 27°, 313°40′ E (S, shell), 46°, 49° (N, nucleus), 46° 139° (S, nucleus), counterclockwise relative orbits, nuclear period 1,080 years, shell 2,273 years	longitude
1757	Leonard Euler (Switzerland)	disjointed dipole N-pole: colat. 14°53′, 250° E (from London?), S-pole: colat. 29°23′, 52°18′ E from N-pole	
1760	Tobias Mayer (Germany)	eccentric precessing(?) dipole colat. 11°30′, increasing yearly by 8′15″, and moving through the interior Earth	longitude?
1766	Richard Lovett (England)	precessing dipole N-pole: colat. 13°51′, 0° longit. from Canaries in 1660, counterclockwise orbit, 506 years	longitude

1777	Daniel Bernoulli (Switzerland)	irregular, isogonic-isoclinic grid	longitude
1778	Le Monnier (France)	disjointed dipole colat. 10°, longit. from Canaries N-pole: 340° E, S-pole: 165° E	
1783	Josef de Porras y Ruiz (Spain)	tilted dipole colat. 26°13′, longit. of Azores	longitude
1787	Tiberius Cavallo (England)	disjointed dipole	
——	John Churchman (Pennsylvania, U.S.A.)	disjointed dipole see table 4.1	longitude
1788	W. Graham (Scotland)	precessing dipole	longitude
——	G. L. Le Clerc, Comte de Buffon (France)	crustal disjointed quadrupole, molten core	longitude
1790	John Churchman (Pennsylvania, U.S.A.)	disjointed dipole see table 4.1	longitude
1793	A. M. Mackay (England)	irregular, latitude-isogonic-isoclinic grid	latitude, longitude
1794	Ralph Walker (England)	disjointed dipole N-pole: colat. 19°, longit. 280° E of London in 1794, S-pole: colat. 25°, longit. 130° E, counterclockwise, irregular orbit	longitude
——	J. Lorimer (England)	disjointed dipole	
——	John Tullock (Scotland)	irregular, isogonic-isoclinic grid	longitude
——	John Churchman (Pennsylvania, U.S.A.)	disjointed dipole see table 4.1	longitude

Notes

O N E : The Earth's Magnetic Field

1. Jack Jacobs, David Gubbins, and Kathy Whaler, interview by author, tape recording, VIIth Geomagnetic Retreat, Kilnsey (Yorkshire), 24–28 Sept. 2001.

2. W. Lowrie, *Fundamentals of Geophysics* (Cambridge, 1997), 178–88; P. Olson, "Probing Earth's Dynamo," *Nature* 389 (1997): 337–38; D. Gubbins, "Geomagnetism: The Next Millennium," *Palaeogeography, Palaeoclimatology, Palaeoecology* 89 (1990): 255, 257–60.

3. D. Gubbins and J. Bloxham, "Morphology of the Geomagnetic Field and Implications for the Geodynamo," *Nature* 325 (1987): 509; C. M. R. Fowler, *The Solid Earth: An Introduction to Global Geophysics* (Cambridge, 1992), 258–59.

4. J. Bloxham and D. Gubbins, "The Secular Variation of Earth's Magnetic Field," *Nature* 317 (1985): 777; W. Kuang and J. Bloxham, "An Earth-like Numerical Dynamo Model," *Nature* 389 (1997): 371.

5. Gubbins, interview.

6. J. Bloxham, D. Gubbins, and A. Jackson, "Geomagnetic Secular Variation," *Philosophical Transactions of the Royal Society* A 329 (1989): 417; J. Bloxham and D. Gubbins, "The Evolution of the Earth's Magnetic Field," *Scientific American* 261 (1989): 68.

7. Jacobs, interview.

8. Whaler, interview.

9. Gubbins, interview.

10. Jacobs, interview.

11. Whaler, interview.

12. Whaler, interview.

13. Whaler, interview.

14. Fowler, *Solid Earth*, 33–36.

15. Gubbins, interview.

16. Bloxham and Gubbins, "Evolution," 70–72; D. Gubbins, "Historical Secular Variation and Geomagnetic Theory," in *Geomagnetism and Palaeomagnetism*, ed. F. J. Lowes et al. (Amsterdam, 1989), 33.

17. P. J. Kelly, "The Time-Averaged Palaeomagnetic Field and Secular Variation" (Ph.D. diss., Univ. of Leeds, 1996); C. L. Johnson and C. G. Constable, "The Time-Averaged Geomagnetic Field: Global and Regional Biases for 0–5 Ma," *Geophysical Journal International* 131 (1997); D. Gubbins and P. J. Kelly, "Persistent Patterns in the Geomagnetic Field over the Past 2.5 Myr," *Nature* 365 (1993); Gubbins and Bloxham, "Morphology," 509; Bloxham, Gubbins, and Jackson, "Geomagnetic Secular Variation."

18. T. Lay, Q. Williams, and E. J. Garnero, "The Core-Mantle Boundary Layer and Deep Earth Dynamics," *Nature* 392 (1998); R. D. van der Hilst, S. Widiyantoro, and E. R. Engdahl, "Evidence for Deep Mantle Circulation from Global Tomography," *Nature* 386 (1997); D. Gubbins and M. Richards, "Coupling of the Core Dynamo and Mantle: Thermal or Topographic?" *Geophysical Research Letters* 13 (1986); R. Hide, "On the Earth's Core-Mantle Interface," *Quarterly Journal of the Royal Meteorological Society* 96 no. 410 (1970): 587.

19. Gubbins, "Historical Secular Variation," 33; Bloxham and Gubbins, "Secular Variation," 778.

20. Gubbins and Bloxham, "Morphology," 511; D. Gubbins and G. Sarson, "Geomagnetic Field Morphologies from a Kinematic Dynamo Model," *Nature* 368 (1994): 55; Johnson and Constable, "Time-Averaged Geomagnetic Field," 664; Gubbins, "Geomagnetism," 257.

21. R. L. Parker, *Geophysical Inverse Theory* (Princeton, N.J., 1994), 296; R. A. Langel, "The Main Field," in *Geomagnetism*, ed. J. A. Jacobs, vol. 1 (London, 1987), 250; Lowrie, *Fundamentals*, 258–63.

22. Gubbins, interview.

23. Bloxham and Gubbins, "Evolution," 71–73; Gubbins, "Geomagnetism," 257–58; Gubbins, "Historical Secular Variation," 33–34; Bloxham and Gubbins, "Secular Variation," 778.

24. J. Bloxham, "Models of the Magnetic Field at the Core-Mantle Boundary for 1715, 1777, and 1842," *J. Geophys. Res.* 91 no. B14 (1986); K. A. Hutcheson and D. Gubbins, "Earth's Magnetic Field in the Seventeenth Century," *J. Geophys. Res.* 95 no. B7 (1990); R. Hide, "Free Hydromagnetic Oscillation of the Earth's Core and the Theory of the Geomagnetic Secular Variation," *Phil. Trans.* A 259 (1966) 615–47; Hide, "Core-Mantle Interface," 579; Bloxham, Gubbins, and Jackson, "Geomagnetic Secular Variation," 467; Langel, "Main Field," 457.

25. The actual resolving power of empirical core-field maps is, however, still being debated; K. Whaler, pers. comm. 22 Oct. 2001.

26. Gubbins, interview.

27. Whaler, interview.

28. Langel, "Main Field," 249.

29. Lowrie, *Fundamentals*, 253–56.

30. S. R. C. Malin, "Historical Introduction to Geomagnetism," in *Geomagnetism*, ed. J. A. Jacobs, vol. 1 (London, 1987), 9–12.

31. Gubbins, interview.

T W O : The Age of Diversity: Geomagnetism before 1600

1. "Of the First Meridian," BL, Sloane MSS 3143 (after 1703), fol. 65r.

2. C. A. Davids, *Zeewezen en Wetenschap: De Wetenschap en de Ontwikkeling van de Navigatietechniek in Nederland tussen 1585 en 1815* (Amsterdam, 1986), 69, 73; A. H. Cook, *Edmond Halley: Charting the Heavens and the Seas* (Oxford, 1998), 397.

3. E. Harrison, *Idea Longitudinis: Being a Brief Definition of the Best Known Axioms for Finding the Longitude . . .* (London, 1696), 76; A. J. Turner, "In the Wake of the Act, but Mainly Before," in *The Quest for Longitude*, ed. W. J. H. Andrewes (Cambridge,

Mass., 1996), 120; F. Marguet, *Histoire de la Longitude à la Mer au XVIIIe Siècle en France* (Paris, 1917), 48.

4. E. G. Forbes, *The Birth of Scientific Navigation: The Solving in the Eighteenth Century of the Problem of Finding Longitude at Sea*, NMM Maritime Monographs and Reports 10 (London, 1974), 2.

5. Moore to Board of Longitude (hereafter BOL), 4 July 1790, CUL, RGO/14/42, no. 7, fol. 148r.

6. C. M. Anhaltin, *Slot en Sleutel van de Navigatie, ofte Groote Zeevaert* (Amsterdam, 1659), 49.

7. Note that if the observer travels at a higher latitude than that of the dipole, the maximum would be 180° declination, found everywhere on the dipole's meridian at a higher latitude.

8. M. Blackman, "The Lodestone: A Survey of the History and the Physics," *Contemporary Physics* 24 no. 4 (1983): 319; J. Daujat, *Origines et Formation de la Théorie des Phénomènes Électriques et Magnétiques*, 3 vols., Exposés d'Histoire et Philosophie des Sciences 989–91 (Paris, 1945), 12, 16, 25, 43; A. Still, *Soul of Lodestone: The Background of Magnetical Science* (New York, 1946), 17–20.

9. D. H. D. Roller, *The "De Magnete" of William Gilbert* (Amsterdam, 1959), 17–18; Still, *Soul of Lodestone*, 13.

10. Still, *Soul of Lodestone*, 14–15; P. Benjamin, *The Intellectual Rise in Electricity: A History* (New York, 1895), 40.

11. H. Balmer, *Beiträge zur Geschichte der Erkenntnis des Erdmagnetismus*, 3 vols., Veröffentlichungen der Schweizerischen Geselschaft für die Geschichte der Medizin und Naturwissenschaft 20 (Aarau, 1956), 170–71; Benjamin, *Intellectual Rise*, 47–49; Roller, *Magnete*, 18–19; Still, *Soul of Lodestone*, 21.

12. A sixteenth-century example echoing this belief is M. Coignet, *Instruction Nouvelle . . . Touchant l'Art de Nauiger . . .* (Antwerp, 1581), 9.

13. Daujat, *Origines*, 12; Benjamin, *Intellectual Rise*, 33, 93; Still, *Soul of Lodestone*, 11; Roller, *Magnete*, 19–20, 24.

14. D. J. Struik, *The Land of Stevin and Huygens: A Sketch of Science and Technology in the Dutch Republic during the Golden Century*, Studies in the History of Modern Science 7 (Dordrecht, 1981), 9–11; P. Radelet-de Grave, *Les lignes Magnétiques de XIIIème Siècle au Milieu du XVIIIème Siècle*, Cahiers d'Histoire et de Philosophie des Sciences, new series, no. 1 (Paris, 1981), 10; Benjamin, *Intellectual Rise*, 90–91; Daujat, *Origines*, 49, 56, 58, 65–66, 88.

15. J. L. Heilbron, *Electricity in the Seventeenth and Eighteenth Centuries: A Study of Early Modern Physics* (Berkeley, Calif., 1979), 23–26; J. A. Smith, "Precursors to Peregrinus: The Early History of Magnetism and the Mariner's Compass in Europe," *Journal of Medieval History* 18 (1992): 36, 62–63.

16. Benjamin, *Intellectual Rise*, 123–25; Roller, *Magnete*, 37–38; for William Gilbert's conclusion to that effect, see A. C. Crombie, *Styles of Scientific Thinking in the European Tradition: The History of Argument and Explanation, Especially in the Mathematical and Biomedical Sciences and Arts*, 3 vols. (London, 1994), 634; G. Sarton, *Six Wings: Men of Science in the Renaissance* (London, 1957), 96.

17. C. A. Ronan, *The Cambridge Illustrated History of the World's Science* (London, 1983), 315; H. L. Hitchins and W. E. May, *From Lodestone to Gyro-compass* (London, 1955), 22; E. O. von Lippmann, *Geschichte der Magnetnadel bis zur Erfindung des Kom-*

passes (gegen 1300), Quellen und Studien zur Geschichte der Naturwissenschaften und der Medizin, vol. 3, Heft 1 (Berlin, 1932), 27–28; Balmer, *Beiträge*, 56; Smith, "Precursors to Peregrinus," 68–69; G. Hellmann, *Rara Magnetica 1269-1599*, Neudrucke von Schriften und Karten über Meteorologie und Erdmagnetismus no. 10 (1898; reprint, Nendeln, 1969), 8; Benjamin, *Intellectual Rise*, 165–78; P. Radelet-de Grave and D. Speiser, "Le 'De Magnete' de Pierre de Maricourt: Traduction et Commentaire," *Revue d'Histoire des Sciences et Leurs Applications* 28 no. 3 (1975): 194.

18. A. E. Nordenskiöld, *Periplus: Essay on Early History of Charts and Sailing Directions*, trans. F. A. Bather (Stockholm, 1897), 49; F. R. Maddison, *Medieval Scientific Instruments and the Development of Navigational Instruments in the XVth and XVIth Centuries*, Agrupamento de Estudos de Cartografia Antiga 30 (Coimbra, 1969), 16; W. E. May, "The Birth of the Compass," *Journal of the Institute of Navigation* 2 (1949): 261; W. D. Hackmann, "Jan van der Straet (Stradanus) and the Origins of the Mariner's Compass," in *Learning, Language, and Invention: Essays Presented to Francis Maddison*, ed. W. D. Hackmann and A. J. Turner (Aldershot, 1994), 158, 162; Benjamin, *Intellectual Rise*, 181; Hellmann, *Rara Magnetica*, 39; for a reconstruction diagram, see J.-M. Martinez-Hidalgo y Teran, *Historia y Leyenda de la Aguja Magnetica: Contribución de los Españoles al Progreso de la Nautica* (Barcelona, 1946), 35.

19. Balmer, *Beiträge*, 62–63; Benjamin, *Intellectual Rise*, 170–78; Radelet-de Grave, *Lignes Magnétiques*, 16–17, 21–22; A. C. Crombie, *Augustine to Galileo: The History of Science AD 400-1650* (London, 1952), 89; Daujat, *Origines*, 86; Hellmann, *Rara Magnetica*, 9–10; Still, *Soul of Lodestone*, 33; Roller, *De Magnete*, 40; Radelet-de Grave and Speiser, *Magnete*, 218–20.

20. L. de Morais e Sousa, *A Sciencia Nautica dos Pilotos Portugueses nos Seculos XV e XVI*, 2 vols. (Lisbon, 1924), 167.

21. Stanley of Alderley, ed., *The First Voyage round the World, by Magellan. Translated From the Accounts of Pigafetta, and Other Contemporary Writers*, Hakluyt Society Publications, 1st series, vol. 52 (London, 1874), 170; note that a one-on-one relationship of degrees of longitude and declination is impossible even in the tilted-dipole case.

22. H. C. Freiesleben, *Geschichte der Navigation* (Wiesbaden, 1978), 43; Smith, "Precursors to Peregrinus," 46–47, 64.

23. E. G. R. Taylor, "Jean Rotz and the Variation of the Compass 1542," *J. Inst. Nav.* 7 (1954): 9; E. G. R. Taylor, *The Haven-finding Art* (London, 1956), 101; Benjamin, *Intellectual Rise*, 176.

24. Nordenskiöld, *Periplus*, 49; Benjamin, *Intellectual Rise*, 96–102, 155–56; Balmer, *Beiträge*, 114, 525–33; Hackmann, "Jan van der Straet," 174; Lippmann, *Geschichte der Magnetnadel*, 11; Still, *Soul of Lodestone*, 58; Roller, *Magnete*, 30, 37.

25. Daujat, *Origines*, 88; Smith, "Precursors to Peregrinus," 52.

26. R. Collinson, ed., *The Three Voyages of Martin Frobisher in Search of a Passage to Cathaia and India by the North-West A.D. 1576-8*, Hakluyt Society Publications, 1st series, vol. 38 (New York, 1867), 34 n. 1; Taylor, *Haven-Finding Art*, 155; A. H. Markham, ed., *The Voyages and Works of John Davis the Navigator*, Hakluyt Society Publications, 1st series, vol. 59 (London, 1880), 342; N. Carpenter, *Geographie Delineated Forth in Two Bookes . . .* (Oxford, 1635), 61.

27. S. P. Thompson, *William Gilbert and Terrestrial Magnetism in the Time of Queen Elizabeth: A Discourse* (London, 1903), 5–6; Benjamin, *Intellectual Rise*, 101–2, 203–4; Still, *Soul of Lodestone*, 57; Balmer, *Beiträge*, 115, 534, 537; Daujat, *Origines*, 111; A. de

Smet, "Gerard Mercator: Zijn Kaarten, Zijn Belangstelling voor het Aardmagnetisme en de Zeevaartkunde," *Mededelingen van de Marine Academie van België* 14 (1962): 129; E. G. R. Taylor, "John Dee and the Map of North-East Asia," *Imago Mundi* 12 (1955): 105.

28. Smet, "Mercator," 131–33; the letter was dated 23 Feb. 1547; an agonic meridian would have to be placed 12° more westerly, so the mountain would lie in 180° E, on the same great circle.

29. Thompson, *William Gilbert*, 5; A. Meskens, "Mercator en de Zeevaart: Enkele Aspecten," in *Gerard Mercator en de Geografie in de Zuidelijke Nederlanden*, Publikaties Museum Plantin-Moretus en Stedelijk Prentenkabinet 29 (Antwerp, 1994), 45–47; D. W. Waters, *The Art of Navigation in England in Elizabethan and Early Stuart Times* (London, 1958), 154; Balmer, *Beiträge*, 124–25, 127, 534–35, 538–42, 544; E. Crone, E. J. Dijksterhuis, and R. J. Forbes, *The Principal Works of Simon Stevin*, vol. 3 (Amsterdam, 1961), 396–99; Smet, "Mercator," 135–37; P. Steenstra, *Openbaare Lessen over het Vinden der Lengte op Zee . . .* (Amsterdam, 1770), 41.

30. Note the similarity with the agonic great circle of a tilted dipole, dividing the globe into two 180° sectors of northwesting and northeasting respectively.

31. C. R. Markham, trans. and ed., *Early Spanish Voyages to the Strait of Magellan*, Hakluyt Society Publications, 2d series, vol. 28 (London, 1911), 13; W. B. Greenlee, trans. and ed., *The Voyage of Pedro Álvarez Cabral to Brazil and India: From Contemporary Documents and Narratives*, Hakluyt Society Publications, 2d series, vol. 81 (London, 1938), xiv; D. C. Goodman, *Power and Penury: Government, Technology, and Science in Philip II's Spain* (Cambridge, 1988), 53, 58–59; Martinez-Hidalgo y Teran, *Historia*, 122; C. R. Boxer, ed., *South China in the Sixteenth Century*, Hakluyt Society Publications, 2d series, vol. 106 (London, 1953), xxxvii–iii; J. P. Sigmond, "De Weg naar de Oost," *Spiegel Historiael* 9 (1974): 360; S. Pumfrey, "'O Tempora, O Magnes!': A Sociological Analysis of the Discovery of Secular Magnetic Variation in 1634," *British Journal for the History of Science* 22 (1989): 190; W. G. L. Randles, "Portuguese and Spanish Attempts to Measure Longitude in the Sixteenth Century," *Vistas in Astronomy* 28 (1985): 236–37.

32. V. M. Godinho, "The Portuguese and the 'Carreira da India' 1497–1810," in *Ships, Sailors, and Spices: East India Companies and Their Shipping in the Sixteenth, Seventeenth, and Eighteenth Centuries*, ed. J. R. Bruijn and F. S. Gaastra (Amsterdam, 1993), 39; Marguet, *Histoire de la Longitude*, 6; Taylor, *Haven-Finding Art*, 185; Hellmann, *Rara Magnetica*, 10–11, 42; L. M. de Albuquerque, *O Livro de Marinharia de André Pires*, Agrupamento de Estudos de Cartografia Antiga vol. 1 (Lisbon, 1963), 5.

33. Fernandez was a captain in the Spanish fleet; Menéndez was governor of Florida and Cuba; Albuquerque, *Livro*, 124; Goodman, *Power and Penury*, 55; G. Hellmann, ed., *Die ältesten Karten der Isogonen, Isoklinen, Isodynamen, 1701, 1721, 1768, 1804, 1825, 1826 . . .* , Neudrucke von Schriften und Karten über Meteorologie und Erdmagnetismus no. 4 (1895, reprint Nendeln 1969), 17; A. Fontoura da Costa, *Ciência Náutica dos Portugueses na Época dos Descobrimentos* (Lisbon, 1958), 66; L. M. de Albuquerque, *Contribuicao das Navegaçoes do Sec. XVI para o Conhecimento do Magnetismo Terrestre*, Agrupamento de Estudos Cartografia Antiga vol. 44 (Coimbra, 1970), 10; J. Keuning, *Petrus Plancius, Theoloog en Geograaf 1552-1622* (Amsterdam, 1946), 124; A. de Santa Cruz, *The Book of Longitudes and Methods That Have Been Kept until Now in the Art of Navigating . . .* , trans. J. Bankston (1542; Bisbee, Ariz., 1992), 24; G. F. de Oviedo, *La Hystoria General de las Indias, Ahora Nueuamente Impressa Corregida y Emendada* (n.p., 1547), xvi; Morais e Sousa, *Sciencia Nautica*, 164; AGI, Indiferente 426, bk. 25 fol. 226r–v.

34. P. de Syria, *Arte de la Verdadera Navegacion* (Valencia, 1602), 58–61; Martinez-Hidalgo y Teran, *Historia*, 99; S. G. Franco, *Historia del Arte y Ciencia de Navegar*, 2 vols. (Madrid, 1947), 62.

35. BL, Regal MSS 20 B VII fol. 16v–19v, 40r, 48r–56r; E. G. R. Taylor, *Tudor Geography 1485-1583* (London, 1930), 63, 65–66; E. G. R. Taylor, "An Elizabethan Compass Maker," *J. Inst. Nav.* 3 (1950): 41; E. G. R. Taylor, "The Dawn of Modern Navigation," *J. Inst. Nav.* 1 (1953): 284; E. G. R. Taylor, *Mathematical Practitioners of Tudor and Stuart England* (Cambridge, 1954), 16; Taylor, *Haven-Finding Art*, 185–86, 189.

36. Toussaints de Bessard, *Dialogue de la Longitude Est-Ouest* (n.p., 1574), 61.

37. Toussaints de Bessard, *Dialogue*, 30, 34–35, 43, 61–62, 65–68; BNP, MSS FR 9175 fol. 3v, 19r–21v; Balmer, *Beiträge*, 114; Marguet, *Histoire de la Longitude*, 62.

38. The main proposal resides in AGI, Patronato 262 R. 4 im. 15, 29–31, 33–40, 52–53, 60, 63, 80–82, 84, 98–103, 116–17; related documents in AGI, Filipinas 340 bk. 3 fol. 35v–36r, 102–5; AGI, Indiferente 428 bk. 33 fol. 115, 125r–v, 167, 175r–v, 178r–v, 184r–185, 188v–189, 191v–192; AGI, Indiferente 428 bk. 34 fol. 3v–4r, 25r; AGI, Patronato 262 R. 3 im. 83, 149–50.

39. AGI, Patronato 262 R. 6 im. 2, 4, 5, 28–30; related correspondence in AGI, Indiferente 428 bk. 34 fol. 181v–182r, 183v–185r, 191r–192v, 202r–v.

40. Both Pulau Batu Putch (1°19′49″ N, 104°24′27″ E, a small island at the eastern entrance of Singapore Strait) and a rock near the Dapeng Bandao Peninsula (at 22°31′59″ N, 114°31′00″ E, very close to Canton in China) were named Pedra Branca by the Portuguese. At both sites has zero declination been observed at the time, rendering geographic attribution ambiguous.

41. Hellmann, *Ältesten Karten*, 16.

42. F. da Costa, "Tratado da Hidrographia y Arte de Navegar" (1596), NMM, MSS NVT/7 fol. 45–49.

43. V. de Saa, *Regimento de Navegaçam* (Lisbon, 1624), 19–21v, 32v; M. de Figueiredo, *Hidrographia Exame de Pilotos . . .* (Lisbon, 1625), 15v–16, 17v.

44. Bruno was also known as Burro, Borro, Borri, and Burri.

45. M. de Figueiredo, *Hydrographie ou Examen, Traduit . . . par Nicolas Le Bon Dieppois, et Augmenté par Luy de la Parfaite Cognoissance de la Variation de l'Aiguille pour Trouuer les Longitudes* (Dieppe, 1640), 23–32.

46. A. de Mariz Carneiro, *Regimento de Pilotos e Roteiro da Navegaçam . . .* (Lisbon[?], 1655), 8, 11, 207; L. S. Pimentel, *Pratica da Arte de Navegar* (1673; facs. reprint, 2d ed., Lisbon, 1960), 56–58; Fontoura da Costa, *Ciência Náutica*, 68; Morais e Sousa, *Sciencia Nautica*, 164, 203; A. Lafuente and M. A. Sellés, "The Problem of Longitude at Sea in the Eighteenth Century in Spain," *Vistas in Astronomy* 28 (1985): 243.

47. A. Pos, "So Weetmen Wat te Vertellen Alsmen Oudt Is: over Ontstaan en Inhoud van het Itinerario," in *Souffrir pour Parvenir: De Wereld van Jan Huygen van Linschoten*, ed. R. van Gelder, J. Parmentier, and V. Roeper (Haarlem 1998), 135, 137, 144–46; J. Parmentier, "In het Kielzog van Van Linschoten: Het 'Itinerario' en het 'Reys-Geschrift' in de Praktijk," ibid. 153; S. P. l'Honoré Naber, *Reizen van Jan Huyghen van Linschoten naar het Noorden, 1594-1595*, Werken der Linschoten Vereniging vol. 8 (The Hague, 1914), xxv–xxvii; P. C. Emmer and F. S. Gaastra, "De Vaart Buiten Europa: het Atlantisch Gebied," in *Maritieme Geschiedenis der Nederlanden*, vol. 2, ed. G. Asaert et al. (Bussum, 1977), 247; Crone, Dijksterhuis, and Forbes, *Principal Works*, 401, 403; Sigmond, "De

Weg naar de Oost," 361; C. R. Boxer, "Portuguese Roteiros 1500–1700," *Mariner's Mirror* 20 (1934): 177–79, 182; Keuning, *Plancius*, 99.

48. G. P. Rouffaer and J. W. IJzerman, *De Eerste Schipvaart der Nederlanders naar Oost-Indië onder Cornelis de Houtman*, 1595–1597, pt. 1, Werken der Linschoten Vereniging vol. 7 (The Hague, 1915), lxxi, lxxiv–v; ibid., pt. 2, Werken der Linschoten Vereniging vol. 25 (The Hague, 1925), 229; G. Schilder and W. F. J. Mörzer Bruyns, "Navigatie," in *Maritieme Geschiedenis der Nederlanden*, ed. G. Asaert et al., vol. 2 (Bussum, 1977), 163; Keuning, *Plancius*, 175; Naber, *Linschoten*, xxix; C. A. Davids, "Finding Longitude at Sea by Magnetic Declination on Dutch East Indiamen, 1596–1795," *American Neptune* 50 no. 4 (1990): 275.

49. The tiny average difference (0.2 degrees) between observed and predicted declinations was obtained at the price of longitudinal adjustments which could exceed ten degrees. For example, Plancius had placed Bantam (on Java) 18°54' west of its true longitude, which fueled a long-standing dispute with van Linschoten on its position.

50. Examples in logbook "Mauritius" (preVOC, 1598), ARA, 1.04.01/43; logbook "Zeelandia" (preVOC, 1598), ARA, 1.04.01/53; logbook "Vriesland" (preVOC, 1598), ARA, 1.04.01/60; logbook "Amsterdam" (preVOC, 1598), ARA, 1.04.01/46; logbook "Utrecht," sailing direction at the back, ARA, 1.04.01/62; logbook "Amsterdam" (VOC, 1613), BL, Egerton MSS 1851, experiencing gross discrepancies between predicted and observed longitudes: at 16 Aug. 1613 the measured declination was found to be off the scale of the longitude-finder; logbook "Nieuw Amsterdam" (VOC), 6 Feb. 1636, UBL, BPL 127 E.

51. J. Hondius, *Tractaet ofte Handelinge van het Gebruyck der Hemelsche en Aertschen Globe* (Amsterdam, 1597), 50–51; A. Haeyen, *Een Corte Onderrichtinge Belanghende die Kunst vander Zeevaert* (Amsterdam, 1600), 10, 16; W. J. Blaeu, *Licht der Zeevaart . . .* (Amsterdam, 1608), preface; R. Robbertsz, *Numeratio: Het Eerste ABC der Tal-Konst* (n.p., 1612), preface, 16, backpage; A. Metius, *Nieuwe Geographische Onderwysinghe . . .* (Franeker, 1614), 10–11, 43–44; Davids, *Zeewezen en Wetenschap*, 76–77, 285–87, 313, 355; Davids, "Finding Longitude," 284, 287; Crone, Dijksterhuis, and Forbes, *Principal Works*, 407–9; Keuning, *Plancius*, 131, 133–35; E. Crone, "Het Aandeel van Simon Stevin in de Ontwikkeling van de Zeevaartkunde," *Mededelingen van de Marine Academie* 15 (1963): 10; S. P. l'Honoré Naber, *Reizen van Willem Barents, Jacob van Heemskerck, Jan Cornelisz. Rijp en Anderen naar het Noorden 1594-1597*, 2 vols., Werken der Linschoten Vereniging vol. 14-15 (The Hague, 1917), xxii–iii, xxix–xxx, 51.

52. S. Stevin, *De Havenvinding* (Leyden, 1599); S. Stevin, *Wisconstighe Ghedachtenissen* (Leyden, 1605-8), (1608) pt. 1 bk. 5, 164–74; Struik, *Stevin and Huygens*, 58; Davids, *Zeewezen en Wetenschap*, 72, 287; E. J. Dijksterhuis, *Simon Stevin* (The Hague, 1943), 186 n. 5, 187–88; Davids, "Finding Longitude," 283–84; Balmer, *Beiträge*, 128, 130–31; Crone, Dijksterhuis, and Forbes, *Principal Works*, 253–54; 365, 368–72, 379; E. J. Dijksterhuis, *Simon Stevin: Science in the Netherlands around 1600* (The Hague, 1970), 77, 83, 91.

53. Thompson, *William Gilbert*, 7; Sarton, *Six Wings*, 94; Taylor, *Tudor and Stuart England*, 47; Waters, *Art of Navigation*, 229; E. Crone, "Het Vinden van de Weg over Zee van Praktijk tot Wetenschap," *Mededelingen van de Koninklijke Vlaamsche Academie voor Wetenschappen . . . van België* 25 no. 9 (1963): 15; Crone, Dijksterhuis, and Forbes, *Principal Works*, 377; Balmer, *Beiträge*, 133.

54. S. Fyler, *Longitudinis Inventae Explicatio Non Longa* ... (London, 1699), preface.

55. T. Blundeville, *The Theoriques of the Seuen Planets, Shewing All Their Diuerse Motions* ... (London, 1602), 288–91; W. Barlowe, *A Briefe Discovery of the Idle Animadversions of Marke Ridley* ... (London, 1618), The English Experience vol. 429 (Amsterdam, 1972), 12.

56. R. Norman, *The Newe Attractiue, Containyng a Short Discourse of the Magnes or Lodestone* ... (London, 1581), 96–97; Hellmann, *Rara Magnetica*, 87.

57. Norman, *Newe Attratiue*, 97–98; W. Barlowe, *Magneticall Advertisements* ... *Concerning the Nature and Properties of the Loadstone* (London, 1616), 86; E. Wright, *Certaine Errors of Navigation Detected and Corrected* ..., 3d ed. (London, 1657), 84; Blundeville, *Theoriques*, 294; B. Keteltas, *Het Ghebruyck der Naeld-wiisinge tot Dienste der Zee-vaert Beschreven* (Amsterdam, 1609), app.; M. Ridley, *A Short Treatise of Magneticall Bodies and Motions* (London, 1613), 131–32; Carpenter, *Geographie Delineated*, 252; J. Holwarda, *Friesche Sterrekonst, ofte een Korte doch Volmaeckte Astronomia* ... (Harlingen, 1652), 276–77; A. Kircher, *Magnes sive de Arte Magnetica* ..., 3d ed. (Rome, 1654), bk. 2, 306–7.

58. G. de Nautonier, *Mecometrie de l'Eymant* ... (Venes, 1603), 46–48, 68.

59. D. Dounot de Barleduc, *Confutation de l'Invention des Longitudes ou de la Mecometrie de l'Eymant* (Paris, 1611), 1–3, 11–12, 26, 36; BODL, Rawlinson papers A 20 (Thurloe papers) no. 16 fol. 332–34; M. Greengrass and M. P. Leslie, eds., *Samuel Hartlib: The Complete Edition* (Ann Arbor, Mich., 1995), cd-rom, 29/4/32b, 71/16/6a.

60. See Hudson's voyages in 1607 and 1608, Feuillée's Atlantic/American survey in the early 1700s, Cook's Pacific ventures in the 1760s and 1770s, and Phipps's Arctic exploits in 1773; Waters, *Art of Navigation*, 271; G. M. Asher, ed., *Henry Hudson the Navigator* ..., Hakluyt Society Publications, 1st series, vol. 27 (London, 1860), 23–34; E. G. R. Taylor, *Mathematical Practitioners of Hanoverian England 1714–1840* (Cambridge, 1966), 54–55, 62; a Danish naval inclination survey off Greenland in logbook "Blaa Heyeren" (SE), 1736, RAK, SE 85.b 1+2/377.

THREE: The Age of Discord: Geomagnetism in the Seventeenth Century

1. Sarton, *Six Wings*, 95; the full title of Gilbert's work is "On the Magnet and Magnetic Bodies, and on That Great Magnet the Earth" (*De Magnete, Magneticisque Corporibus, et de Magno Magnete Tellure*).

2. The term *effluvium* can denote many things; Gilbert saw it as a sort of radiation (actually comparing it to light (in bk. 2, ch. 7), exciting a struck object's magnetic potential, but not remaining as a constant medium; Crombie, *Styles*, 634; Daujat, *Origines*, 134; J. A. Bennett, "Cosmology and the Magnetic Philosophy (1640–80)," *Journal for the History of Astronomy* 12 (1981): 166.

3. D. Goodman and C. A. Russell, eds., *The Rise of Scientific Europe, 1500–1800* (Sevenoaks, 1991), 200–201; Thompson, *William Gilbert*, 13–14; Benjamin, *Intellectual Rise*, 19, 279–92; Roller, *De Magnete*, 128–31, 138–53; Crombie, *Augustine to Galileo*, 319, 321, 342; Crombie, *Styles*, 632–34; Still, *Soul of Lodestone*, 100–102, 118–19.

4. P. Dear, *Discipline and Experience: The Mathematical Way in the Scientific Revolution* (Chicago, 1995), 65–66, 158–60; S. Lilley, *The Development of Scientific Instruments in the Seventeenth Century* (London, 1951), 66; A. R. Hall, *The Revolution in Science, 1500-1750* (London, 1983), 257; Heilbron, *Electricity*, 171–73; Crombie, *Styles*,

633–34; Benjamin, *Intellectual Rise*, 272, 276–77, 291–92; Roller, *De Magnete*, 94, 128–29, 134, 137, 153; Balmer, *Beiträge* 155; Daujat, *Origines*, 125–35; W. Gilbert, *De Magnete*, trans. P. Fleury Mottelay (New York, 1958), 233, 243; Thompson, *William Gilbert*, 9; Waters, *Art of Navigation*, 246.

5. Gilbert, *De Magnete*, 251–54.

6. Gilbert, *De Magnete*, 253–54; Sarton, *Six Wings*, 97; Gilbert's cosmological views were dispersed in English in Blundeville, *Theoriques of the Seuen Planets*.

7. Dijksterhuis, *Science*, 70–71, 76; Benjamin, *Intellectual Rise*, 163; Still, *Soul of Lodestone*, 91.

8. Benjamin, *Intellectual Rise*, 284, 292–93; Roller, *De Magnete*, 48, 146, 162–64, 171–73; ironically, it is in fact due to tidal forces that the Moon on average rotates with the exact speed required to continually show earthlings the same side.

9. ARA, 1.04.02/1133 fol. 84r–v; J. B. Cohen, *Revolution in Science* (Cambridge, Mass., 1985), 134; Daujat, *Origines*, 164; M. R. Baldwin, "Magnetism and the Anti-Copernican Polemic," *J. Hist. Astr.* 16 no. 3 (1985): 155; Benjamin, *Intellectual Rise*, 267–70, 273, 327.

10. Baldwin, "Magnetism," 231; R. G. Collingwood, *The Idea of Nature* (Oxford, 1945), 101–2; Crombie, *Augustine to Galileo*, 321; Balmer, *Beiträge*, 137–40, 160–62, 403–4, 414–15, 421–22, 425; Daujat, *Origines*, 155, 166, 169, 178, 252–53; Roller, *De Magnete*, 173, 183–84; Crombie, *Styles*, 615; Benjamin, *Intellectual Rise*, 261, 344–45; Still, *Soul of Lodestone*, 89.

11. T. Browne, *Pseudodoxia Epidemica: or Enquiries into Very Many Received Tenets and Commonly Presumed Truths* (London, 1646), 63–64; Benjamin, *Intellectual Rise*, 335–36; Ridley, *Magneticall Bodies and Motions*, 96–103; Barlowe, *Magneticall Advertisements*, 49–51; J. Verqualje, *Het Tweede Boeckjen, Verklarende Wat Dit . . . Son-compas Is, ende . . . om daer door Oost en West . . . te Konnen Bezeylen* (Amsterdam, 1661), 46–47; G. Denys, *L'Art de Naviger, Perfectionné par la Connaissance de la Variation de l'Aimant . . .* (Dieppe, 1666), 27–28.

12. A. G. Debus, *Man and Nature in the Renaissance* (Cambridge, 1978), 105; Collingwood, *Idea of Nature*, 101; D. H. D. Roller, "Did Bacon Know Gilbert's 'De Magnete'?," *Isis* 44 (1953): 13; Sarton, *Six Wings*, 97; Benjamin, *Intellectual Rise*, 323–26; Still, *Soul of Lodestone*, 94; Daujat, *Origines*, 154; Waters, *Art of Navigation*, 282; L. Reael, *Observatien of Ondervindinge aen de Magneetsteen, en de Magnetische Kracht der Aerde* (Amsterdam, 1651), 78–81.

13. Baldwin, "Magnetism," 155, 160, 166–68; Heilbron, *Electricity*, 10–13; Daujat, *Origines*, 157.

14. N. Cabeo, *Philosophia Magnetica in qua Magnetis Natura Penitus Explicatur . . .* (Ferrara, 1629), 91, 218; Baldwin, "Magnetism," 156–58; Dear, *Discipline and Experience*, 65; Heilbron, *Electricity*, 28, 180–81; Still, *Soul of Lodestone*, 168; Daujat, *Origines*, 193–94, 200–201, 203, 206–9, 216; Benjamin, *Intellectual Rise*, 349–54.

15. Kircher, *Magnes*, 295, 312–28, 334–37, 346–47; Baldwin, "Magnetism," 158–60; compare present-day metaphorical use of "veins" of ore.

16. J. Grandamy, *Nova Demonstratio Immobilitatis Terrae Petita ex Virtute Magnetica . . .* (La Flèche, 1645), preface.

17. Baldwin, "Magnetism," 167–68, 171–72; Daujat, *Origines*, 183; Benjamin, *Intellectual Rise*, 405.

18. Grandamy, *Nova Demonstratio*, 73, 75–79, 81, 83.

19. I.-B. Riccioli, *Geographiae et Hydrographiae Reformatae* . . . , 12 vols. (Venice, 1672), 351.

20. S. R. C. Malin and E. C. Bullard, "The Direction of the Earth's Magnetic Field at London (1570–1975)," *Phil. Trans.* A 299 (1981): 384.

21. E. Gunter, *De Sectore et Radio* (London, 1623).

22. M. Christy, ed., *The Voyages of Captain Luke Foxe of Hull and Captain Thomas James of Bristol, in Search of a North-West Passage in 1631-32* . . . , 2 vols., Hakluyt Society Publications, 1st series, vol. 88–89 (London, 1894), 613.

23. H. Gellibrand, *A Discourse Mathematical on the Variation of the Magneticall Needle* (London, 1635), Neudrucke von Schriften und Karten über Meteorologie und Erdmagnetismus no. 9, ed. G. Hellmann (1897; reprint Nendeln, 1969), 19; S. F. Mason, *A History of the Sciences* (New York, 1962), 253; A. McConnel, *Geomagnetic Instruments before 1900: An Illustrated Account of Their Construction and Use* (London, 1980), 5; M. Feingold, *The Mathematicians' Apprenticeship: Science, Universities, and Society in England, 1560-1640* (Cambridge, 1984), 11; C. H. Cotter, "Edmund Gunter (1581-1626)," *J. Inst. Nav.* 34 (1981): 367; Goodman and Russell, *Scientific Europe*, 206–7; Malin, "Historical Introduction," 19–20; Sarton, *Six Wings*, 94; Taylor, *Tudor and Stuart England*, 38, 62–63, 72–73; Taylor, *Haven-finding Art*, 232; Benjamin, *Intellectual Rise*, 447; Crombie, *Styles*, 636; G. Fournier, *Hydrographie Contenant la Théorie et la Pratique de Toutes les Parties de la Navigation*, 2d ed. (Paris, 1676), 415–16; Flamsteed to Pepys, 21 Apr. 1697, in E. G. Forbes, L. Murdin, and F. Willmoth, eds., *The Correspondence of John Flamsteed, the First Astronomer Royal*, 2 vols. (Bristol, 1995, 1997), 2:633.

24. Skepticism in F. Jacobssen [Visscher], *Discours opt Nuttelijck Gebruijck van de Peijlcompassen* (Batavia, 1642), fol. 511v–12r, and in Verqualje, *Tweede Boeckjen*, 11–12; tables without expiry date in Anhaltin, *Slot en Sleutel*, 49, 56–59, and in D. R. van Nierop, *Nieroper Schatkamer, Waermee dat de Kunst der Stuerluyden door Seeckere Grontregulen Geleert en Gebruickt Kan Worden* (Amsterdam, 1676), 59–60.

25. A. C. Hellingwerf, *Hoornse Beknopte Stuurmanskunst* (Hoorn, 1694), 336–82; P. van Musschenbroeck, *Physicae Experimentales et Geometricae de Magnete* . . . (Ph.D. thesis, University of Leyden, 1729), 150–88; F. W. Stapel and C. W. Th. Baron van Boetzelaer van Asperen en Dubbeldam, *Pieter van Dam. Beschryvinge van de Oostindische Compagnie*, 7 vols., Rijks Geschiedkundige Publikatiën vol. 63, 68, 74, 76, 83, 87, 96 (The Hague, 1976), bk. 1, 2:681; Davids, *Zeewezen en Wetenschap*, 77–80; Davids, *Finding Longitude*, 288.

26. BL, Add. MSS 4393 no. 4 fol. 37v; Pumfrey, *O Tempora*, 207–8.

27. Pumfrey, *O Tempora*, 208–13; Balmer, *Beiträge*, 175; Denys, *Art de Naviger*, 55.

28. Holwarda, *Friesche Sterrekonst*, 354–55.

29. P. Petit, "An Extract of a Letter Sent from Paris about the Loadstone . . . ," *Phil. Trans.* 2 no. 28 (1667): 531; Auzout, "Observation," 1187; H. S. Jones, "The Early Years of the Royal Society," *J. Inst. Nav.* 13 (1960): 377; P. Fleury Mottelay, *Bibliographical History of Electricity and Magnetism Chronologically Arranged* (London, 1922), 130.

30. "Manuscrits Delisle" (ca.1705-17), ANP, MAR 2JJ/59; "Mémoires Delisle" (1710), BNP, MSS NAF 10764; ANP, MAR 3JJ/36 (after 1720).

31. N. Sarrabat, *Nouvelle Hipothese sur les Variations de l'Eguille Aimantée* . . . (Bordeaux, 1727), 1–9.

32. P. Bouguer, *Nouveau Traité de Navigation, Contenant la Théorie et la Pratique de Pilotage* (Paris, 1753), 313–16; Académie Royale des Sciences, *Table Générale des Matières*

Contenues dans l'Histoire et les Mémoires de l'Académie Royale des Sciences de Paris . . . , 7 vols. (Amsterdam, 1791), 4:414; Bézout, *Suite du Cours de Mathématiques . . . Contenant le Traité de Navigation* (Paris, 1775), 54; Le Gaigneur, *Le Pilote Instruit, ou Nouvelles Leçons de Navigation Sans Maitre . . .* (Nantes, 1781), 458.

33. S. Shapin, *The Scientific Revolution* (Chicago, 1996), 47; Radelet-de Grave, *Lignes Magnétiques,* 30; Heilbron, *Electricity,* 31; E. Marcorini, ed., *The History of Science and Technology: A Narrative Chronology,* 2 vols. (New York, 1988), 1:188–89; Benjamin, *Intellectual Rise,* 359–61; Daujat, *Origines,* 285.

34. Daujat, *Origines,* 298–302, 308, 311; Benjamin, *Intellectual Rise,* 357–61; Still, *Soul of Lodestone,* 114–15, 164–65.

35. Forbes, Murdin, and Willmoth, *Correspondence of John Flamsteed,* 1:923; Balmer, *Beiträge,* 224; Still, *Soul of Lodestone,* 120, 144; J. Churchman, *The Magnetic Atlas, or Variation Charts of the Whole Terraquous Globe . . . ,* 4th ed. (London, 1804), viii–ix; Baldwin, "Magnetism" 158; Daujat, *Origines,* 285, 321–22, 324, 329–30, 333, 339–40; S. Pumfrey, "Mechanizing Magnetism in Restoration England: the Decline of Magnetic Philosophy," *Annals of Science* 44 (1987): 4.

36. Daujat, *Origines,* 251, 375–76, 380, 382, 388, 391–95.

37. Benjamin, *Intellectual Rise,* 376–79; Still, *Soul of Lodestone,* 116–17.

38. Feingold, *Mathematicians' Apprenticeship,* 5, 8, 188–89, 214, 216; J. A. Bennett, *The Divided Circle: A History of Instruments for Astronomy, Navigation, and Surveying* (Oxford, 1987), 54–55; Goodman and Russell, *Scientific Europe,* 83; Ronan, *World's Science,* 373–74.

39. Radelet-de Grave, *Lignes Magnétiques,* 37; Benjamin, *Intellectual Rise,* 379–80, 404, 406, 413, 445–46.

40. H. Power, *Experimental Philosophy in Three Books, Containing New Experiments Microscopical, Mercurial, Magnetical . . .* (London, 1664), 155–56, 159; Still, *Soul of Lodestone,* 140; Benjamin, *Intellectual Rise,* 414–17; Daujat, *Origines,* 255–56, 261, 265–66.

41. C. G. Dyer, "Newton: A Man of His Times," *J. Inst. Nav.* 28 (1975): 210–11; Heilbron, *Electricity,* 38; Pumfrey, *Mechanizing Magnetism,* 1–4; Bennett, *Divided Circle,* 51; Still, *Soul of Lodestone,* 141.

42. Pumfrey, *Mechanizing Magnetism,* 1–21.

43. R. W. Home, "'Newtonianism' and the Theory of the Magnet," *Hist. Sci* 15 (1977): 255–60; Benjamin, *Intellectual Rise,* 436–38; Cook, *Halley,* 176; Radelet-de Grave, *Lignes Magnétiques,* 34; Dyer, "Newton," 209; Fleury Mottelay, *Bibliographical History,* 134; Pumfrey, *Mechanizing Magnetism,* 16; P. Fara, *Sympathetic Attractions: Magnetic Practices, Beliefs, and Symbolism in Eighteenth-Century England* (Princeton, N.J., 1996), 16, 178–79; Daujat, *Origines,* 401; Dijksterhuis, *Science,* 1.

44. R. K. Merton, *Science, Technology, and Society in Seventeenth-Century England* (New York, 1970), 163–64, 175–76; L. S. Feuer, *The Scientific Intellectual: The Psychological and Sociological Origins of Modern Science* (New York, 1963), 52; Marcorini, *History of Science,* 1:191; Fara, *Sympathetic Attractions,* 31, 35–36; Feingold, *Mathematicians' Apprenticeship,* 188; T. Birch, *The History of the Royal Society of London,* 4 vols. (London, 1756, 1757), 2:163.

45. Greengrass and Leslie, *Hartlib,* 71/16/1a.

46. Greengrass and Leslie, *Hartlib,* 71/16/1a.

47. Birch, *History,* 2:319; Taylor, *Tudor and Stuart England,* 91; Greengrass and Leslie, *Hartlib* 8/49/1a–2b, 28/1/8b.

48. H. Bond, *The Longitude Found: or, a Treatise Shewing an Easie and Speedy Way . . . to Find the Longitude . . .* (London, 1676), 7; the 1666 scheme is detailed in BL, Sloane MSS 3964 fol. 1–16.

49. Birch, *History,* 1:104; [C. Cavendish], "The Variations of the Magnetick Needle Predicted for Many Yeares Following," *Phil. Trans.* 3 no. 40 (1668): 789–90; Bond's table listed predictions for 1663 and 1669–1716.

50. "The Undertaking of Mr Henry Bond Senior . . . Concerning the Variation of the Magnetical Compass . . . ," *Phil. Trans.* vol. 8 no. 95 (1673): 6065–66; BL, Add. MSS 4393 no. 4 fol. 37v–38.

51. Birch, *History,* 3:130–31.

52. Birch, *History,* 3:467; *The Posthumous Works of Robert Hooke . . . ,* ed. R. Waller (London, 1705), 484–86; Ronan, *World's Science,* 90.

53. BL, Add. MSS 4393 no. 2 fol. 8–10; ibid. no. 4 fol. 26–47v.

54. BL, Add. MSS 4393 no. 4 fol. 79, 89, 91; Pumfrey, *Mechanizing Magnetism,* 15; Taylor, *Haven-Finding Art,* 247; Taylor, *Tudor and Stuart England,* 90, 111–12; Taylor, "Modern Navigation," 287.

55. Bond, *Longitude Found,* 6–7; E. Halley, "A Proposal for Finding the Longitude at Sea within a Degree, or Twenty Leagues," *Phil. Trans.* 37 (1731): 186.

56. RS, CP 9 no. 31 fol. 90; Birch, *History,* 3:336.

57. P. Blackborrow, *The Longitude Not Found: or, an Answer to a Treatise, Written by Henry Bond . . .* (London, 1678), preface, 2, 10–45, 61–77; J. Seller, *Practical Navigation or an Introduction to the Whole Art* (London, 1689), 155, reproduced the prediction table for London (1689–1716); a later edition, revised by John Colson (1730) introduced the same table by stating "that the motion of the variation is slowest when it is nearest its period of greatest deviation according to the following Theory of Mr Bond" (127); J. Atkinson, *Epitome of the Art of Navigation; or, a Short and Easy Methodical Way to Become a Compleat Navigator* (London, 1707), 254; ibid., revised and ed. W. Mountaine (London, 1744, 1753, 1757, 1767); ibid., revised and ed. J. Adams (London, 1790).

58. H. Wallis, "Navigators and Mathematical Practitioners in Samuel Pepys's Day," *J. Inst. Nav.* 47 (1994): 4; E. G. R. Taylor, *Navigation in the Days of Captain Cook,* Maritime Monographs and Reports 18 (London, 1975), 7; Taylor, *Haven-finding Art,* 250–53, 259; C. H. Cotter, "Captain Edmond Halley RN, FRS," *Notes and Records of the Royal Society* 36 no. 1 (1981): 70; Forbes, Murdin, and Willmoth, *Correspondence of John Flamsteed,* 1:620, 922, 2:1037, 1045.

59. Flamsteed to Moore, 30 Apr. 1678, in Forbes, Murdin, and Willmoth, *Correspondence of John Flamsteed,* 1:620; Flamsteed to Towneley, 22 Oct. 1678 (on a lunar eclipse), ibid. 1:652; Flamsteed to Halley, 26 Nov. 1679 (on a mathematical problem), ibid., 1:716; ibid. 2:1045; Birch, *History,* 4:7.

60. Flamsteed to Towneley, 13 Feb. 1680, RS, MSS 243 fol. 45v; Forbes, Murdin, and Willmoth, *Correspondence of John Flamsteed,* 1:737–38, n. 6; memorandum Flamsteed, 12 Dec. 1700, ibid. 2:876; Birch, *History,* 4:6, 15, 17–19, 21–22.

61. Birch, *History,* 4:19.

62. Birch, *History,* 4:21, 24, 26; Flamsteed to Towneley, 15 Dec. 1680, in Forbes, Murdin, and Willmoth, *Correspondence of John Flamsteed,* 1:747–48; Flamsteed to Molyneux, 25 Sept. 1682, ibid. 2:44–46; Taylor, *Tudor and Stuart England,* 116.

63. Birch, *History,* 4:206.

64. E. Halley, "A Theory of the Variation of the Magnetical Compass," *Phil. Trans.* 13 no. 148 (1683): 216, 219.

65. Halley, "Theory," 220.

66. Halley, "Theory," 221; the text was reiterated in Royal Society, *Miscellania Curiosa: Containing a Collection of Some of the Principal Phenomena in Nature . . .* , 3 vols., 2d ed. (London, 1708), 1:27–42.

67. Other factors include Flamsteed's resentment over Halley's disagreement regarding Flamsteed's explanations of the tides at Dublin, the conviction that Halley's associates were spreading tales that Flamsteed should vacate his position as Astronomer Royal in favor of Halley, and the complaint of having entertained Halley hospitably at Greenwich, but being treated meanly himself when he stayed at Halley's overnight; Cook, *Halley*, 131, 174–75, 380, 382.

68. Flamsteed to Molyneux, 12 Oct. 1681, in Forbes, Murdin, and Willmoth, *Correspondence of John Flamsteed*, 1:828.

69. Flamsteed to Towneley, 4 Nov. 1686, in Forbes, Murdin, and Willmoth, *Correspondence of John Flamsteed*, 2:298.

70. Memorandum Flamsteed, 2 Feb. 1695, in Forbes, Murdin, and Willmoth, *Correspondence of John Flamsteed*, 2:563–64; Cook, *Halley*, 175.

71. Flamsteed to Thomas Perkins, 11 Dec. 1700, in Forbes, Murdin, and Willmoth, *Correspondence of John Flamsteed*, 2:874.

72. Memorandum Flamsteed, 12 Dec. 1700, in Forbes, Murdin, and Willmoth, *Correspondence of John Flamsteed*, 2:875; letter Flamsteed to Pound, 13 Dec. 1700, ibid. 2:878; Flamsteed to Wren, 19 Nov. 1702, ibid. 2:982; E. Halley, "Proposal," 192.

73. BODL, Rigaud MSS 37 fol. 43, 61; other comparative experiments on the force of lodestones in Birch, *History*, 4:526.

74. E. Halley, "An Account of the Cause of the Change of the Variation of the Magnetical Needle, with an Hypothesis of the Structure of the Internal Parts of the Earth," *Phil. Trans.* 17 no. 195 (1692): 565.

75. Halley, "Hypothesis," 567.

76. Halley, "Hypothesis," 567–70; in RS, CP 21 no. 37 (around 1692), Halley postulated a nuclear circular motion about twice as fast.

77. Cook, *Halley*, 147–78, 191–92, 417.

78. W. Whiston, *The Longitude and Latitude Found by the Inclinatory or Dipping Needle; wherein the Laws of Magnetism Are Also Discover'd . . .* (London, 1721), 78; Euler to the Princess of Anhalt-Dessau, 24 Oct. 1761, in *Lettres de L. Euler à une Princesse d'Allemagne sur Divers Sujets de Physique et de Philosophie*, ed. A. A. Cournot, 2 vols. (Paris, 1842), 261.

79. BODL, Rigaud MSS 37 fol. 133; J. B. Scarella, *De Magnete Libri Quatuor*, 2 vols. (Bressanone, 1759), 2:10–75, 176–206; Semeyns figures in the next chapter.

F O U R: The Age of Data: Geomagnetism in the Eighteenth Century

1. M. Alexandrescu, V. Courtillot, and J.-L. le Mouël, "Geomagnetic Field Direction in Paris since the Mid–Sixteenth Century," *Physics of the Earth and Planetary Interiors* 98 (1996): 326–38; Marin Mersenne to Theodore Haack, 25 Feb. 1640, 4 Mar. 1640, Greengrass and Leslie, *Hartlib*, 18/2/18a–19a; "Observatio Declinationis Magneticae Parisiis" (1660), ibid. 62/49/3a.

2. "Journal des Observations" (1734–35), ANP, MAR 3JJ/47 no. 5; Esprit Pézenas to Jacques Cassini, 3 Jan. 1735, ANP, MAR G/92 fol. 53.

3. Académie Royale des Sciences, *Table Générale*, 1:30–42, 4:7, 82, 5:418.

4. N. Cruquius, *Tegenwoordige Miswysing der Compassen rondom den Aardbol . . .* (Haarlem, 1738); "Registers Spaarndam" (1705–37, 1745–58), HHRL, cat. no. 11069; M. Semeyns, *Kortbondige Demonstratio . . . om door Twee Vaste en Vier Beweegende Punten van Trekkinge . . . de Wyzing der Compassen te Bereekenen . . .* (The Hague, 1762); D. J. Slikker, *Klaar Bewys over het Onmogelyk der Oost en Westvinding . . .* (Amsterdam, 1703), 50–53; Musschenbroeck, *Physicae*, 151–234.

5. BL, Add. MSS 38823 fol. 2–5; D. B. Quinn, ed., *The Roanoke Voyages 1584–1590: Documents to Illustrate the English Voyages to North America . . .* , 2 vols., Hakluyt Society Publications, 2d series, vol. 104 (London, 1955), 51–53; BL, Sloane MSS 427; Greengrass and Leslie, *Hartlib*, 71/16/4a–5b.

6. Birch, *History*, 1:20, 2:288, 4:35, 163, 435, 448; Halley to Hans Sloane, 12 Oct. 1696, BODL, Rigaud MSS 37, fol. 200–201; Robert Hooke to Herming of Altorf, 5 Apr. 1680, BL, Sloane MSS 1039 fol. 173.

7. Birch, *History*, 2:177, 288, 4:489; RS, JBC 8, fol. 303; BODL, Rigaud MSS 37 fol. 92, 103; Whiston, *Laws of Magnetism*, vii; Malin and Bullard, "London 1570–1975," 360, 370–72, 390, 416.

8. BL, Add. MSS 4436 no. 19 fol. 128; "Directions for Seamen, Bound for Far Voyages," *Phil. Trans.* 1 no. 8 (1665): 140–42; [H. Oldenburg], "Directions for Observations and Experiments to Be Made by Masters of Ships, Pilots, and Other Fit Persons in Their Sea-Voyages," *Phil. Trans.* 2 no. 24 (1667): 433–48; Radelet-de Grave, *Lignes Magnétiques*, 30–40; Mason, *History of the Sciences*, 269; Flamsteed to Pepys, 21 Apr. 1697, in Forbes, Murdin, and Willmoth, *Correspondence of John Flamsteed*, 2:633.

9. RS, JBC 8, fol. 164; C. A. Ronan, *Edmond Halley, Genius in Eclipse* (London, 1970), 161.

10. N. J. W. Thrower, ed., *The Three Voyages of Edmond Halley in the "Paramore" 1698–1701*, 2 vols., Hakluyt Society Publications, 2d series, vol. 156 (London, 1980), 18–20, 30; Cook, *Halley*, 262, 264–65, 290.

11. The expedition lasted from 20 October 1698 to 22 June 1699; during the voyage, Harrison had made his own magnetic observations, which differed from the captain's; after his court case, the lieutenant decided to offer his work to Flamsteed (known foe of Halley for reasons earlier outlined), who copied Harrison's journal in March 1700; Forbes, Murdin, and Willmoth, *Correspondence of John Flamsteed*, 2:783–84, 811–13, 1040; Flamsteed to Thomas Burchett, 23 June 1699, 4 July 1699, in E. F. MacPike, ed., *Correspondence and Papers of Edmond Halley* (London, 1932), 107–9, 245.

12. Thrower, *Voyages of Edmond Halley*, 35–48; RS, JBC 9, fol. 171; BODL, Rigaud MSS 37 fol. 82; the 1699 chart was not an isogonic representation but a regular one, with plotted values of declination.

13. Birch, *History*, 3:467; BODL, Rigaud MSS 37, fol. 91–92; Whiston, *Laws of Magnetism*, xxii–iv, 42, 104, 111; R. Walker, *A Treatise on Magnetism, with a Description and Explanation of a Meridional and Azimuth compass . . .* (London, 1794), 165–92; Le Monnier, *Loix du Magnétisme, Comparées aux Observations et aux Expériences . . .* (Paris, 1776), xxvii, 61, 121; ANP, MAR 3JJ/47 no. 13; Le Monnier to French Naval Ministry, 11 May 1777, ANP, MAR G/95 fol. 48–50; ANP, MAR G/99 fol. 86; Forbes, Murdin, and Willmoth, *Correspondence of John Flamsteed*, 2:1046; Hellmann, *Ältesten Karten*, 13; Taylor,

Hanoverian England, 54–55; W. F. J. Mörzer Bruyns, "Longitude in the Context of Navigation," in *The Quest for Longitude*, ed. W. J. H. Andrewes (Cambridge, Mass., 1996), 59.

14. J. Michell, "Experiments on Two Dipping Needles," *Phil. Trans.* 62 (1772): 476–80; R. Waddington, *Epitome of Theoretical and Practical Navigation . . .* (London, 1777), 20–21; M. Basso Ricci and P. Tucci, "Considerazione Storiche e Didattiche su Strumenti e Methodi di Misura del Magnetismo Terrestre," *Musescienza* 4/5 (1993): 31; Taylor, *Hanoverian England*, 62–63; Alexandrescu, Courtillot, and Le Mouël, "Geomagnetic Field Direction," 339–48.

15. G. Graham, "An Account of the Observations Made of the Horizontal Needle at London, in . . . 1722," *Phil. Trans.* 33 (1724): 96; RS, CP 9, pt. 2 no. 18, fol. 33; McConnel, *Geomagnetic Instruments*, 5–6; Malin, "Historical Introduction," 20–22.

16. RS, CP 4, pt. 2, no. 6, fol. 4, 5, 15–18, 20–97; J. Canton, "An Attempt to Account for the Regular Diurnal Variation of the Horizontal Magnetic Needle; and also . . . at the Time of an Aurora Borealis," *Phil. Trans.* 51 pt. 1 (1759): 398–445; Malin, "Historical Introduction," 22; Still, *Soul of Lodestone*, 120, 126.

17. "Mémoire sur les Variations des Aiguilles Aimantées, et sur les Boussoles," ANP, MAR G/99 fol. 89–92; Heilbron, *Electricity*, 75; McConnel, *Geomagnetic Instruments*, 9.

18. E. Halley, "An Account of the Late Surprizing Appearance of the Lights Seen in the Air . . . , with an Attempt to Explain the Principal Phaènomena Therof," *Phil. Trans.* 29 no. 347 (1716): 423.

19. Halley, "Appearance of the Lights," 406–8, 421–28; A. Armitage, *Edmond Halley* (London, 1966), 182; note that Halley had previously stated that the medium was a non-magnetic fluid; another postulate of life in the interposing medium was put forward by Whiston, *Laws of Magnetism*, 75: "Perhaps therfore that fluid may prove such as is fit for animals to live in; I mean either on the inner surface of the upper Earth, or outward surface of the central loadstone, or else in the very fluid itself also; which last is the case of the upper surface of our Earth, and of the fluid air above us."

20. BODL, Rigaud MSS 37 fol. 133; Malin, "Historical Introduction," 22–23, 45; McConnel, *Geomagnetic Instruments*, 6; Canton, "Diurnal Variation," 399–403.

21. Daujat, *Origines*, 398–99, 420, 477, 483; Still, *Soul of Lodestone*, 42, 166; BL, Add. MSS 4433 fol. 200; Radelet-de Grave, *Lignes Magnétiques*, 32; [G.-A. Bazin], *Description des Courants Magnetiques . . . Suivie de Quelques Observations sur l'Aiman* (Strasbourg, 1753), 41.

22. Sarrabat, *Nouvelle Hipothese*, 10–12, 14–21, 24–26, 35–36, 40–41, 43, 45–48; Marcorini, *History of Science*, 1:174, 241, 260; note that the longitudinal distance between the Azores and Canton actually comes up short of 180 degrees.

23. BL, Add. MSS 4436 no. 19 fol. 136v; Daujat, *Origines*, 399, 449, 452, 459–60; Heilbron, *Electricity*, 92–95; Marguet, *Histoire de la Longitude*, 58; Radelet-de Grave, *Lignes Magnétiques*, 97.

24. Home, "Newtonianism," 253–55, 258–63; T. S. Kuhn, "Mathematical versus Experimental Traditions in the Development of Physical Science," in *Natuurwetenschappen van Renaissance tot Darwin*, ed. H. A. M. Snelders and K. van Berkel (The Hague, 1981), 42.

25. Cruquius notebooks (1725, 1727, 1730), HHRL, no. 11069; Cruquius, *Tegenwoordige Miswysing*; A. F. van Engelen and H. A. M. van Geurts, *Nicolaus Cruquius (1678-1754) and His Meteorological Observations*, KNMI Historische Weerkundige Waarnemingen 4 (De Bilt, 1985), iv–v, 15, 18, 20, 27–28, 151; A. R. T. Jonkers, "Finding

Longitude at Sea: Early Attempts in Dutch Navigation," *De Zeventiende Eeuw* 12 no. 1 (Hilversum, 1996), 193.

26. BNP, MSS NAF 10764; ANP, MAR 3JJ/7 no. 17.

27. ANP, MAR G/91 no. 8 fol. 22; ibid. no. 1; ANP, MAR 3JJ/47 no. 6.

28. ANP, MAR G/91 nos. 6–8; ANP, MAR 3JJ/7 no. 26; Delacroix to De Maurepas, 3 June 1746, ANP, MAR G/91 no. 8 fol. 35–36; J.-Ph. de la Croix, *Plan Abregé de la Découverte des Longitudes tant sur Mer que sur Terre* (Paris, 1747), 1–4.

29. BL, Add. MSS 4433 fol. 67v.

30. BL, Add. MSS 4433 fol. 64r.

31. BL, Add. MSS 4433 fol. 62v, 66v, 77, 78–83, 85–86.

32. Samuel Johnson (a.k.a. Dr. Johnson, 1709–84) was an English poet, essayist, critic, journalist, lexicographer, and conversationalist, and wrote a *Dictionary of the English Language* (1755), the first major dictionary to use illustrative historical quotations.

33. Z. Williams, *The Mariner's Compass Compleated: or, the Expert Seaman's Best Guide* (London, 1745); Z. Williams, *A True Narrative of Certain Circumstances Relating to Zachariah Williams . . .* (London, 1749); Z. Williams, *An Account of an Attempt to Ascertain the Longitude at Sea by an Exact Theory of the Variation of the Magnetical Needle* (London, 1755); RS, JBC 13, fol. 323; ibid. JBC 17, fol. 99, 428.

34. E. G. Forbes, ed., *The Theory of the Magnet and Its Application to Terrestrial Magnetism*, vol. 3 of *The Unpublished Writings of Tobias Mayer*, Niedersächsische Staats-und Universitätsbibliothek Göttingen, Arbeiten 11 (Göttingen, 1972), 7–9, 11–13, 23, 99–104; E. G. Forbes, "Tobias Mayer (1723–62): A Case of Forgotten Genius," *Brit. J. His. Sci.* 5 (1970): 17; Heilbron, *Electricity*, 91; Fleury Mottelay, *Bibliographical History*, 220.

35. A. Maasdorp, *Vindinge der Lengtens van den Gehelen Aardbodem . . .* (Amsterdam, 1753), 4, 20–24, 29–30.

36. Whiston, *Laws of Magnetism*, 53–55.

37. E. Swedenborg, *Principia Rerum Naturalium sive Novorum Tentaminum Phaenomena Mundi Elementaris Philosophice Explicandi*, vol. 1 of *Opera Philosophica et Mineralia* (Dresden, 1734), 312–13.

38. Euler had been a student of the Bernoulli's, and was a neo-Cartesian in assuming the magnetic fluid to move from the equator to the poles; Fleury Mottelay, *Bibliographical History*, 213; Still, *Soul of Lodestone*, 170–71; Radelet-de Grave, *Lignes Magnétiques*, 32, 40, 103–5; Euler, *Lettres*, 235, 251–52, 255–56, 261.

39. Semeyns, *Demonstratio*; ARA, 1.02.04/8481; GAD, MSS 38 D 20; D. Bierens de Haan, "Een Leidsch Hoogleraar en een Enkhuizer Natuurkundige in de Vorige Eeuw," *Album der Natuur* (1872): 244–45; a reference to Semeyns's system in a navigational textbook in A. Struick, *Verhandeling over de Zeevaartkunde* (Rotterdam, 1768), 306 n.

40. G.-L. Le Clerc, Comte de Buffon, *Histoire Naturelle des Minéraux*, 5 vols. (Paris, 1783–88), 70, 79–82, 88–92; the author deemed deposits of iron ore to affect the global field only once exposed to the air, when they could partake in atmospheric circulation; their general pattern formed two streams, from the equator to each pole.

41. Walker, *Treatise on Magnetism*, 6–7, 44, 196–97.

42. Walker, *Treatise on Magnetism*, 11–12, 14, 17, 29, 32–35, 39, 44–45, 198; despite the advertised temporary validity of his system, Walker nevertheless claimed the longitude reward of £10,000 (assuming an accuracy to within sixty nautical miles); Walker to BOL, 6 Dec. 1794, CUL, RGO/14/29 no. 8 fol. 108v–10v.

43. At 39°43′ north latitude, extending five degrees west longitude from Delaware; CUL, RGO/14/11 no. 11 fol. 97.

44. CUL, RGO/14/42 no. 5 fol. 60–63, 65–66, 71–72, 74–77.

45. CUL, RGO/14/42 no. 5 fol. 115, 132.

46. CUL, RGO/14/11 no. 11 fol. 97–98, 103–4; ibid. no. 5 fol. 133–34.

47. J. Churchman, *An Explanation of the Magnetic Atlas or Variation Chart . . .* (Philadelphia, 1790); RGO/14/42 no. 5 fol. 78–116.

48. CUL, RGO/14/42 no. 5 fol. 117–19.

49. CUL, RGO/14/42 no. 5 fol. 121, 123; J. Churchman, *The Magnetic Atlas or Variation Charts of the Whole Terraqueous Globe . . .*, 2d ed. (London, 1794).

50. CUL, RGO/14/11 no. 11 fol. 101; J. Churchman, *The Magnetic Atlas or Variation Charts of the Whole Terraqueous Globe . . .*, 3d ed. (New York, 1800); Churchman, *Magnetic Atlas*, 4th ed.

51. CUL, RGO/14/42 no. 5 fol. 135–36.

52. CUL, RGO/14/42 no. 5 fol. 138.

53. S. Silverman, "A Forgotten Proposal for Determination of the Temporal Variation of the Magnetic Declination," *Eos: Transactions American Geophysical Union* 79 (1998): 305.

F I V E : Traversing the Trackless Oceans

1. E. G. R. Taylor, ed., *A Regiment for the Sea and Other Writings on Navigation by William Bourne of Gravesend, a Gunner (ca. 1535-1582)*, Hakluyt Society Publications, 2d series, vol. 121 (Cambridge, 1963), 168–69.

2. De Tourville, "Le Petit Manuel de Pilote" (early eighteenth century), BNP, FR MSS 22046 fol. 118, 120v, 123v–4r; more on these "clues" in chapter 7; Waddington, *Epitome*, illustrated the difference between coastal and oceanic navigation by remarking that "all ships of consequence, especially those bound beyond the equator, take with them azimuth compasses" to observe magnetic declination with (19).

3. Examples of Dutch East Indiamen ending up in the doldrums in logbook "Berckenroode" (VOC, 1701), ARA, 1.04.02/5122, and logbook "Huys Overrijp" (VOC, 1701), ARA, 1.04.02/5124.

4. The passage was marked on some Dutch charts with two lines delineating the longitudinal boundaries, their similarity to two ruts causing it to be dubbed the "Cart Track" (*wagenspoor*).

5. Extreme examples in logbook "Zuijd Beveland" (VOC), 21 Dec. 1771, RAZ, 33.1/187 (7° more westerly than estimated on sighting the Cape of Good Hope), and logbook "Onslow" (EIC), 25 Sept. 1745, IOBL, L/MAR/B 164 E (after surviving the doldrums and ending up on the Angolan coast, when realizing an earlier island sighting had been incorrectly identified).

6. An exception to the rule was the *Lively*, which completed the voyage from Bengal to England in 142 days, without a single stopover; logbook "Lively" (EIC, 1783–84), IOBL, L/MAR/B 534 A.

7. C. R. Markham, *A Life of John Davis, the Navigator, 1550-1605, Discoverer of Davis Straits* (London, 1889), 146; W. C. Dampier, *A History of Science and Its Relations with Philosophy and Religion*, 4th ed. (Cambridge, 1948), 100; Struik, *Stevin and Huygens*, 27–28; M. W. Richey, "The Haven-Finding Art," *J. Inst. Nav.* 10 (1957): 275.

8. Logbook "Tyger Prize" (RN), 28 July 1691, PRO, ADM 52/112; more examples in logbook "Prins Willem de Vijfde" (MCC, 1759), RAZ, 20/989; logbook "Mercurius" (MCC), 1 July 1758, RAZ, 20/758; logbook "Zuijd Beveland" (VOC), 21 Dec. 1771, RAZ, MSS 33.1/187; logbook "London" (EIC), 12 May 1761, IOBL, L/MAR/B 600 A.

9. F. Dassié, *Le Routier des Indes Orientales et Occidentales Traitant des Saisons Propres à y Faire Voyage . . .* (Paris, 1677), 70.

10. Logbook "Ceres" (ADM), 29 Apr. 1789, NSM, cat.I/303; d'Après de Mannevillette, "Mémoire sur la Navigation de France aux Indes," *Mémoirs de Mathematiques et de Physique Presentés à l'Académie des Sciences* (Paris, 1768), 209–10 n.

11. "Instructie voor de Retourschepen" (repr. 15 Sept. 1769), ARA, 1.04.02/3250 fol. 583; "Carte de l'Europe," in *Le Petit Atlas Maritime: Recueil de Cartes et Plans des Quatre Parties du Monde*, vol. 4, comp./ed. J.-N. Bellin (Paris, 1764), no. 1; examples of Dutch charts in French logbooks in logbook "Jeux" (CDI, 1689), ANP, MAR 4JJ/90; logbook "Charente" (FAD), 25 Dec. 1732, ANP, MAR 4JJ/44; logbook "Sirene" (FAD), 27 May 1750, ANP, MAR 4JJ/43; an example of a French chart used on board a Dutch vessel in logbook "Castor" (ADM), 8 June 1770, ARA, 1.01.46/1170.

12. Regulations for handing in complete logbooks or extracts were already formulated by the EIC in 1602 (and reiterated by a 1614 injunction); the RN started collecting them from 1673; eighteenth-century instructions in ARA, 1.04.02/5019 (VOC, after 1746); GAA, PA 231 (Marquette)/223 (4 July 1749) stated the officers would only receive payment after submitting their logbooks; logbook "Comte d'Argenson" (CDI, 1765–67) preface, ANP, MAR 4JJ/92 (limit fifteen days); BL, Add. MSS 38355 (24 Oct. 1799); RAK, AK 194b (n.d.) 24 (no. 41) warned of a penalty of a month's wages on failing to hand in the manuscript.

13. Examples of columnized layout: logbook "Zierikzee" (VOC, 1620), ARA, 1.10.30/309; logbook "Falcon" (EIC, 1625), IOBL, L/MAR/A 42; logbook "Kingfisher" (RN, 1686), PRO, ADM 52/56; logbook "Christiansborg" (GK, 1722–25), RAK, 85.b 1+2/320a; logbook "Grev Laurvig" (SE, 1725–27, printed columns), RAK 85.b 1+2/348d; logbook "Excellent" (FAD, 1680–81), ANP, MAR 4JJ/43; logbook "Royale" (CDI, 1682–83), ANP, MAR 4JJ/144 O; a late VOC example of diary-style is logbook "Zeewijk" (VOC, 1726–28), ARA, 1.04.02/11417; early occurrences of English "block" design in anonymous logbook (EIC, 1662), IOBL, L/MAR/A 70 and logbook "Seahorse" (RN, 1735–38), PRO, ADM 52/484.

14. Examples of printed column format in S. Dunn, *The Longitude Journal; Its Description and Application* (London, 1789); logbook "'s Hertogenbosch" (VOC, 1640–43), NLS, Adv. MSS 33.3/4; logbook "Samuel and Anna" (EIC, 1705), IOBL, L/MAR/A 171; logbook "Victoire" (FAD, 1718–19), ANP, MAR 4JJ/21; logbook "Cavalier" (CDI, 1733–34), ANP, MAR 4JJ/91; logbook "Cron Printzen" (AK, 1734–35), RAK, VA 14/758; in the English merchant marine: logbook "Simon Taylor" (MER, 1784–85), NMM, LOG/M 4; printed block format on HBC ships for instance in logbook "Seahorse" (HBC, 1751), PRO, HB 1/1629; examples of VOC layout copied by other organizations in logbook "Philadelphia" (MCC, 1762–63), RAZ, 20/929: logbook "Utrecht" (ADM, 1783), ARA, 1.10.11.02/40–41; logbook "Kongen af Danmark" (AK, 1738–39), RAK, VA 14/999; logbook "Dronningen af Danmark" (AK, 1742–44) RAK, VA 14/1003.

15. Navigation textbook examples in C. J. Lastman, *De Schatkamer des Grooten Seevaerts-Kunst* (Amsterdam, 1653), 175; Seller, *Practical Navigation*, 348; NLS, MSS 9593 Ewen (1732) fol. 111; H. Wilson, *Navigation New Modell'd, or a Treatise of Geometrical,*

Trigonometrical, Arithmetical, Instrumental, and Practical Navigation, 2d ed. (London, 1723), 205, 208; NMM, MSS NVT/3 De Junquieres (1736) fol. 117–19; J. Barrow, *Navigation Britannica, or a Complete System of Navigation . . .* (London, 1750), 272, 278; T. Haselden, *The Seaman's Daily Assistant, Being a . . . Method of Keeping a Journal at Sea* (London, 1776), 126–60; instructions on logkeeping for instance in RAZ, 103/4 ADM (1744) art. 2; ARA, 1.04.02/5019 (VOC, after 1746); GAA, PA 231/223 (ADM, 1749); ARA, 1.01.46/801a (ADM, 1755) 8, item 21, and ibid. (1793) 16, item 11; S. Dunn, *Rules for a Ship's Journal at Sea* (London, 1784).

16. Logbook "St. Joseph" (FAD, 1711–14), ANP, MAR 4JJ/47 and logbook "Sirene" (CDI, 1720), ANP, MAR 4JJ/90 clearly distinguished observed and estimated magnetic declination, the former by using Arabic numerals for allowed values and Roman for measured ones.

17. Logbook "Chandos" (EIC), 28 Sept.–5 Oct. 1690, IOBL, L/MAR/A 89.

18. C. R. Boxer, ed., *The Tragic History of the Sea 1589-1622,* Hakluyt Society Publications, 2d series, vol. 92 (Cambridge, 1959), 14; A. H. Markham, ed., *The Voyages and Works of John Davis the Navigator,* Hakluyt Society Publications, 1st series, vol. 59 (London, 1880), 240; C. H. Cotter, "Nautical Astronomy: Past, Present, and Future," *J. Inst. Nav.* 29 (1976): 335; H. A. Foreest and A. de Booy, *De Vierde Schipvaart der Nederlanders naar Oost-Indië onder Jacob Wilkens en Jacob van Neck, 1599-1604,* Werken der Linschoten Vereniging vol. 82–83 (The Hague, 1980–81), 46; E. G. R. Taylor, ed., *The Troublesome Voyage of Captain Edward Fenton, 1582-1583,* Hakluyt Society Publications, 2d series, vol. 113 (Cambridge, 1959), 299 n. 1; J. H. Parry, *The Age of Reconnaissance* (London, 1963), 91–92; W. F. J. Mörzer Bruyns, *The Cross-Staff: History and Development of a Navigational Instrument* (Zutphen, 1994); Davids, *Zeewezen en Wetenschap,* 65–67, 120–26, 165–77; Wilson, *Navigation New Modell'd,* 160.

19. P. Collinder, "A New History of Navigation," *J. Inst. Nav.* 9 (1956): 361; Davids, *Zeewezen en Wetenschap,* 275; J. Keuning, *De Tweede Schipvaart der Nederlanders naar Oost-Indië onder Jacob Cornelisz van Neck en Wybrant Warwyck 1598-1600,* vol. 2, Werken der Linschoten Vereniging vol. 44 (The Hague, 1940), xxiii; W. B. Greenlee, *Cabral,* 37 n. 3.

20. C. H. Cotter, *A History of Nautical Astronomy* (London, 1986), 141–43; Sarton, *Six Wings,* 90; L. C. Wroth, *The Way of a Ship: An Essay on the Literature of Navigational Science* (Portland, Me., 1937), 30; Markham, *John Davis,* 341.

21. P. K. Seidelman, P. M. Janiczek, and R. F. Haupt, "The Almanacs: Yesterday, Today, and Tomorrow," *J. Inst. Nav.* 30 (1977): 312; D. W. Waters, "Nautical Astronomy and the Problem of Longitude, in *The Uses of Science in the Age of Newton,* ed. J. G. Burke (Berkeley Calif., 1983), 148–49, 158; Fleury Mottelay, *Bibliographical History,* 220; W. K. Lamb, *George Vancouver: A Voyage of Discovery to the North Pacific Ocean and Round the World, 1791-1795,* 4 vols., Hakluyt Society Publications, 2d series, vol. 163–66 (London, 1984), 49; G. Schilder and W. F. J. Mörzer Bruyns, "Navigatie," in *Maritieme Geschiedenis der Nederlanden,* ed. G. Asaert et al., vol. 3 (Bussum, 1977), 215–16, 222; Marguet, *Histoire de la Longitude,* 6; Davids, *Zeewezen en Wetenschap,* 67, 178–79, 283; McConnel, *Geomagnetic Instruments,* 20; Forbes, "Forgotten Genius," 1–20.

22. Mason, *History of the Sciences,* 270; Struik, *Stevin and Huygens,* 26, 75; Davids, *Zeewezen en Wetenschap,* 67, 87; C. D. Andriesse, *Titan Kan Niet Slapen: een Biografie van Christiaan Huygens* (Amsterdam, 1993), 122, 148–51, 205, 215–17, 269, 303, 325–26, 337–38, 355, 366–70; M. S. Mahoney, "Christiaan Huygens: The Measurement of Time

and of Longitude at Sea," in *Studies on Christiaan Huygens*, ed. H. J. M. Bos et al. (Lisse, 1980), 246–47, 252–60, 264, 269; E. G. Forbes, "The Bicentenary of the Nautical Almanac (1767)," *Brit. J. His. Sci.* 3 (1967): 393; Lamb, *Vancouver*, 50; Waters, "Nautical Astronomy," 150.

23. Bennett, *Divided Circle*, 179; Schilder and Mörzer Bruyns, "Navigatie," 3:213, 215–16, 222–23; C. A. Davids, "De Zeevaartkunde en Enkele Maatschappelijke Veranderingen in Nederland tussen 1850 en 1914," *Meded. Ned. Ver. Zeegesch.* 40–41 (1980): 55; Davids, *Zeewezen en Wetenschap*, 183–84, 216, 283; S. E. Morison, "The Northern Voyages A.D. 500–1600," in *The European Discovery of America*, pt. 1 (New York, 1971), 141; A. Pogo, "Gemma Frisius: His Method of Determining Differences of Longitude by Transporting Time-pieces (1530) . . . ," *Isis* 2 pt. 2 no. 64 (1935): 469; Forbes, "Forgotten Genius," 18–20; D. Howse, "The Astronomers Royal and the Problem of Longitude," *Antiquarian Horology* 21 (1993): 50; J. Sutton, *Lords of the East: The East India Company and Its Ships* (London, 1981), 107; F. Russo, "L'Hydrographie en France aux XVI et XVIII Siècles: Écoles et Ouvrages d'Enseignement," in *Enseignement et Diffusion des Sciences en France au XVIII Siècle*, ed. R. Taton (Paris, 1964), 420.

24. Logbook "Maarseveen" (VOC, 1664), BL, Egerton MSS 1852 no. 2 fol. 125v; logbook "Neptune" (EIC), 21 May 1766, IOBL, L/MAR/B 98 D; logbook "Tryall" (RN), 18 Sept. 1752, PRO, ADM 52/736; W. E. May, "Navigational Accuracy in the Eighteenth Century," *J. Inst. Nav.* 6 (1953): 71.

25. S. Dunn, *Navigator's Guide to the Oriental or Indian Seas or the Description and Use of a Variation Chart of the Magnetic Needle . . .* (London, 1775), 6.

26. Marguet, *Histoire de la Longitude*, 25, 30–31; J. R. Bruijn, F. S. Gaastra, and I. Schöffer, *Dutch-Asiatic Shipping*, vol. 1, Rijks Geschiedkundige Publikatiën 165 (The Hague, 1987), 75, 91; E. Gøbel, "The Danish Asiatic Company's Voyages to China, 1732–1833," *Scand. Econ. Hist. Rev.* 27 (1979): 42–43; P. C. Newman, *Company of Adventurers: The Story of the Hudson's Bay Company*, vol. 1 (Markham, Ontario, 1985), 150; Taylor, *Tudor and Stuart England*, 153; Waters, *Art of Navigation*, 290.

27. Greenlee, *Cabral*, 37 n. 3; E. Crone, "De Zestiende-Eeuwse Zeeman als Kartograaf," *KNAG Geografisch Tijdschrift* 1 (1967) 174–78; C. Denoix, "Les Problèmes de Navigation au Début des Grandes Découvertes," in *Le Navire et l'Economie Maritime du XVe au XVIIIe Siècle*, ed. M. Mollat (Paris, 1957), 135–36; logbook "Neptune" (EIC), 25 May 1766, IOBL, L/MAR/B 98 D; Harrison, *Idea Longitudinis*, 44: "For the log is but a false supposition to find the distance run"; examples of blaming the currents in logbook "Onslow" (EIC), 25 Sept. 1745, IOBL, L/MAR/B 164 E; logbook "Prospect" (EIC), 11 Mar. 1682, NMM, LOG/C/45; logbook "Utrecht" (preVOC), 24 July 1600, ARA, 1.04.01/62; logbook "Zutphen" (VOC), 29 July 1632, ARA, 1.04.02/1105; logbook "Zuijd Beveland" (VOC), 31 Oct. 1771, RAZ, MSS 33.1/187; logbook "Argonaute" (CDI), 14 May 1731, ANP, MAR 4JJ/74; logbook "Duc de Chartres" (CDI), 13 Mar. 1734, ANP, MAR 4JJ/86; logbook "Coche" (CDI), 28 Apr. 1687, ANP, MAR 4JJ/93.

28. [E. Halley], "An Advertisement Necessary for All Navigators Bound up the Channel of England," *Phil. Trans.* 22 (1701): 725–26; similar warnings in Dassié, *Routier*, 10–11; *Atlas Maritimis et Commercialis . . . to Which Are Added Sailing Directions . . .* , comp. J. Knapton and J. Knapton (London, 1728), 14–15; Walker, *Treatise on Magnetism*, 202–4; T. Duckworth, "Directions for Coming into the English Channel" (after 1794), NMM, MSS DUC/45; honest admittance in logbooks of underestimation of magnetic

declination is rare—see logbook "Maarseveen" (VOC), 27 Jan. 1664, BL, Egerton MSS 1852 no. 2.

s i x : Following the Iron Arrow

1. BODL, Tanner Papers 88 no. 25 fol. 245r–v.

2. Absence of awareness of magnetic declination is evident in maps and charts displaying a general tilt in orientation caused by uncorrected declination; BL, Regal MSS 20 B 7 (1542) fol. 19v, 53–55; W. Cunningham, *The Cosmographical Glasse, Conteinyng the Pleasant Principles of Cosmographie, Geographie, Hydrographie, or Nauigation* (London, 1559), 160–61; Taylor, *Tudor Geography*, 26–27; J. de Lisboa, "Tratado da Agulha de Marear" (1514), in *Livro de Marinharia: Roteiros, Sondas e Outros Conhecimentos Relativos á Navegaçao*, ed. J. I. de Brito Rebello (Lisbon, 1903), fol. 9v–11v; F. R. Maddison, *Medieval Scientific Instruments and the Development of Navigational Instruments in the XVth and XVIth Centuries*, Agrupamento de Estudos de Cartografia Antiga 30 (Coimbra, 1969), 47; Godinho, "Carreira da India," 3–5, 25, 35, 39; Albuquerque, *Contribuicao*, 6–7; L. M. de Albuquerque, *Curso de História da Náutica* (Coimbra, 1972), 214–15; Boxer, "Portuguese Roteiros," 172; Hellmann, *Rara Magnetica*, 13; Albuquerque, *Livro*, 7–8.

3. Balmer, *Beiträge*, 85–86, 203; G. Hellmann, "Die Anfänge der Magnetischen Beobachtungen," *Acta Cartographica* 6 (1969): 183–86; Markham, *John Davis*, 344; Goodman, *Power and Penury*, 54, 59; Marguet, *Histoire de la Longitude*, 6; Wroth, *Way of a Ship*, 50, 52; Goodman and Russell, *Scientific Europe*, 119, 123; Taylor, *Haven-Finding Art*, 182, 185; Crone, "De Weg over Zee," 11; Hellmann, *Rara Magnetica*, 11–13; Albuquerque, *Livro*, 5; E. G. R. Taylor, ed., *Roger Barlow, A Brief Summe of Geographie*, Hakluyt Society Publications, 2d series, vol. 69 (London, 1932), xvii; Struik, *Stevin and Huygens*, 42; Taylor, *Tudor Geography*, 83; Boxer, "Portuguese Roteiros," 177, 182; Boxer, *History of the Sea*, 13; Keuning, *Plancius*, 141–42; Morais e Sousa, *Sciencia Nautica*, 205–8; Albuquerque, *Contribuicao*, 11; Fontoura da Costa, *Ciência Náutica*, 69–70.

4. P. de Medina, *The Arte of Nauigation, Wherein Is Contained All the Rules . . . Which for Good Nauigation Are Necessarie . . .*, trans. J. Frampton (1545, London, 1581) ch. 3, 4, 6.

5. D. Garcia de Palacio, *Nautical Instruction for the Good Use and Management of Ships . . .* (Mexico, 1587), trans. J. M. Bankston (Bisbee, Ariz., 1987), bk. 1, ch. 9; Markham, *John Davis*, 145; Balmer, *Beiträge*, 86, 403–13; Toussaints de Bessard, *Dialogue*, ii–iii, 23; Coignet, *Instruction*, 12; R. Norman, *Newe Attractiue*, 90, 102; C. R. Markham, ed., *Tractatus de Globis et Eorum Usu. A Treatise Descriptive of the Globes Constructed by Emery Molyneux, and Published in 1592, by Robert Hues*, Hakluyt Society Publications, 1st series, vol. 79 (London, 1889), 119–20; Borough (1596), 165; E. Wright, *Certaine Errors*, 68; A. de Graaf, *De Seven Boecken van de Groote Zeevaert . . .* (Amsterdam, 1658), 45; Anhaltin, *Slot en Sleutel*, 47; Denys, *Art de Naviger*, 53; Riccioli, *Geographiae*, 474; C. F. Millet Dechales, *L'Art de Naviger . . .* (Paris, 1677), preface; Musschenbroeck, *Physicae*, 151; S. Dunn, *Navigator's Guide*, 8; Hellmann, *Rara Magnetica*, 17; H.-G. Körber, *Zur Geschichte der Konstruktion von Sonnenuhren und Kompassen des 16. bis 18. Jahrhunderts*, Veröffentlichungen des Staatlichen Mathematisch-Physikalischen Salons vol. 3 (Berlin, 1965), 66.

6. BL, Sloane MSS 46 A no. 2, fol. 151–52; Dassié, *Routier*, 15; logbook "Prospect" (EIC), 11 Mar. 1682, NMM, LOG/C/45.

7. C. R. Markham, trans./ed., *The Voyages of Pedro Fernandez de Quiros, 1595 to 1606*, 2 vols., Hakluyt Society Publications, 2d series, vol. 14–15 (1904; repr. Nendeln, 1967), 388, 402; Keuning, *Tweede Schipvaart*, xxviii; R. Zamorano, *Cort Onderwiis van de Conste der Seevaert . . .*, trans. M. Everart (Amsterdam, 1598), 31; D. B. Quinn and A. M. Quinn, eds., *The New England Voyages, 1602-1608*, Hakluyt Society Publications, 2d series, vol. 161 (London, 1983), 289; W. Foster, ed., *The Voyage of Nicholas Downton to the East-Indies 1614-15 . . .*, Hakluyt Society Publications, 2d series, vol. 82 (London, 1939), 2–3; M. Christy, ed., *The Voyages of Captain Luke Foxe of Hull and Captain Thomas James of Bristol, in . . . 1631-32 . . .*, 2 vols., Hakluyt Society Publications, 1st series, vol. 88–89 (London, 1894), 210.

8. Logbook "Samaritan" (EIC, 1615), BL, Egerton MSS 2121 fol. 8; logbook "Kempthorne" (EIC, 1669–70), BL, Sloane MSS 3814 fol. 101–2; logbook "Susanna" (EIC, 1683), IOBL, L/MAR/A 72; star azimuths in logbook "Bellona" (ADM), 19 Aug. 1789, MMPH, MSS.

9. Example in logbook "Hollandia" (VOC), 19 Feb. 1625, ARA, 1.04.02/5049.

10. Problems of single amplitude and advantages of azimuth are discussed in "Horlogerie de la Marine," ANP, MAR G/98 fol. 300v; S. le Cordier, *Instruction des Pilotes . . .* (Havre de Grace, 1754), 138–39; Steenstra, *Openbaare Lessen*, 46; C. Pietersz, *Handleiding tot het Practicale of Werkdadige Gedeelte van de Stuurmanskunst* (Amsterdam, 1779), 29–30; observation techniques in R. Baron "Elements of Navigation" (1755), BL, Add. MSS 15239 fol. 240v; d'Après de Mannevillette, *Navigation*, 218 n.; an example of sighting problems due to atmospheric conditions in logbook "Castor" (ADM), 7 and 19 June 1769, ARA, 1.01.46/1170.

11. Harrison, *Idea Longitudinis*, 41; Dunn, *Navigator's Guide*, 15; Pietersz, *Stuurmanskunst*, 125–26.

12. Logbook "Paramore" (RN), 15 Feb. 1699, BL, Add. MSS 30368 fol. 23; early Dutch examples of azimuth include logbook "Wapen van Delft" (ADM), 29–30 Nov. 1623, ARA, SG 1.05.01/9290; logbook "Eenhoorn" (VOC), 23 June 1693, ARA, 1.04.02/5058.

13. Compagnie des Indes: 18.75 percent; French Navy: 10.21 percent.

14. C. R. Markham, ed., *The voyages of William Baffin, 1612-1622*, Hakluyt Society Publications, 1st series, vol. 63 (London, 1881), xlviii–ix; Christy, *North-West Passage*, 607.

15. R. Norwood, *The Seaman's Practice, Contayning a Fundamental Probleme in Navigation . . .*, 16th ed. (1637; London, 1686), 108.

16. Fournier, *Hydrographie*, 123; Wilson, *Navigation New Modell'd*, 163; J. Harris, *A Treatise of Navigation* (London, 1730), 206–7; J. Moore, *A New Systeme of the Mathematicks . . .* (London, 1681), 195; Dunn, *Navigator's Guide*, 7.

17. Dassié, *Routier*, 84; A. Linton, *Newes of the Complement of the Art of Navigation . . .* (London, 1609), 20; a very rare reference to use of the binnacle compass for observation in logbook "Duc d'Anjou" (CDI), 29 May 1735, ANP, MAR 4JJ/114.

18. Early Dutch examples in logbook "Mauritius" (preVOC), 18 June 1598, ARA, 1.04.01/43; logbook "Griffioen" (VOC), 13 Feb., 10 Apr. 1608, ARA, 1.11.01.01/1137; logbook "Swol" (VOC), 1 Aug. 1637, ARA, 1.04.02/1119; eighteenth-century examples in logbook "Princesse Carolina" (ADM), 1 Dec. 1760, ARA, 1.01.46/1157; logbook "Johanna Catriena" (MER), 22 June 1782, GAR, MSS 3312.

19. Christy, *North-West Passage*, 279; logbook "Banda" (VOC), 27 Mar. 1636, ARA, 1.04.02/1120; logbook "Royalle" (CDI), 17 June 1682, ANP, MAR 4JJ/90; logbook "Vosmaar" (VOC), 5 Aug. 1696, ARA, 1.04.02/5063.

20. Examples in logbook "St. Antoine" (FAD?), 9 Oct. 1709, ANP, MAR 4JJ/1; logbook "Catherine" (EIC), 22 Aug. 1710, IOBL, L/MAR/B 115 B; logbook "Royal Duke" (EIC), 31 Mar. 1758, IOBL, L/MAR/B 614 C; logbook "London" (EIC), 22 Apr. 1761, IOBL, L/MAR/B 600 A.

21. P. Bouguer, *De la Méthode d'Observer en Mer la Déclinaison de la Boussole* (Paris, 1731), 32, 41, 46, 50; J. Mudbey, "Remarks on Time-Keepers and Navigation" (1793), RS, MM vol. 7 no. 102 fol. 13–14; logbook "Bellona" (ADM), 12 Apr. 1789, MMPH, MSS.

22. Logbook "Banda" (VOC, 1636–37), ARA, 1.04.02/1120 fol. 189; logbook "Sunderland" (RN), 18 June 1710, PRO, ADM 52/293; logbook "Montagu" (EIC), 3 and 27 Apr. 1745, IOBL, L/MAR/B 552 J; logbook "Warwick" (RN), 26 Aug. 1774, PRO, ADM 52/742; logbook "Bellona" (ADM), 11 Jan. 1792, MMPH, MSS.

23. Logbook "Royalle" (CDI), 22 Oct. 1682, ANP, MAR 4JJ/144 O; Verqualje, *Tweede Boeckjen*, 42.

24. Logbook "Dronningen af Danmark" (AK, 1742–44), RAK, AK 1003; logbook "Flore" (FAD), 29 Feb. 1772, ANP, MAR 4JJ/2; logbook "Cerf" (FAD, 1756), ANP, MAR 4JJ/87; logbook "Brisson" (CDI, 1769), ANP, MAR G/244; logbook "Amsterdam" (VOC), 9 July 1633, ARA, 1.04.02/1110 fol. 530.

25. Logbook "Kingfisher" (RN), 14 June–24 July 1704, PRO, ADM 52/200 speaks of the "Queen's" and "Capt.'s Compass"; logbook "Beschermer" (ADM), 17–24 Jan. 1738, ARA, 1.01.47.17/43 (after which date the naval issue compass was no longer used); logbook "Glory" (RN), 12 Jan.–21 Apr. 1752, PRO, ADM 52/867; logbook "Yarmouth" (RN), 10–14 Aug. 1760, PRO, ADM 52/1113; logbook "Panther" (RN), 25 Feb.–11 Mar. 1760, PRO, ADM 52/991; logbook "Huijs ter Duijne" (VOC), 12 Aug.–29 Oct. 1696, ARA, 1.04.02/5061; logbook "Eenhoorn" (VOC), 23 July–4 Sept. 1698, ARA, 1.04.02/5075; logbook "Goidschalkoord" (VOC), 11 Apr.–19 July 1737, ARA, 1.04.02/6033; logbook "Brederode" (ADM), 25 July–8 Sept. 1741, ARA, 1.01.46/2338; logbook "Herculis" (VOC), 6 Aug–23 Oct. 1750, GAA, M 165; logbook "Polanen" (VOC), 7 Feb.–8 May 1750, ARA, 1.11.01.01/1304; logbook "Flamand" (FAD), 30 Oct. 1783, ANP, MAR 4JJ/109.

26. G. della Porta, *Natural Magick*, intro. D. J. Price (1658; facs. repr., Washington, D.C., 1957), 211–22.

27. Porta, *Natural Magick*, 212.

28. W. Barlowe, *The Navigator's Supply* (1597; facs. Amsterdam, 1972), B3r; Benjamin, *Intellectual Rise*, 142; Still, *Soul of Lodestone*, 37; Daujat, *Origines*, 117.

29. Examples in logbook "Maasnimph" (ADM), 7 Oct. 1787, ARA, 1.01.46/1215; logbook "Howland" (EIC), 14, 26 Apr. 1704, IOBL, L/MAR/B 696 A; Newman, *Hudson's Bay Company*, 149; C. R. Markham, trans./ed., *Narratives of the Voyages of Pedro Sarmiento de Gamboa to the Straits of Magellan*, Hakluyt Society Publications, 1st series, vol. 91 (London, 1895), 92.

30. J. McCluir, "Continuation of the Description of the Coast of Malabar . . . 1789 and 1790" (1791), NMM, MSS HMN/175 fol. 33.

31. R. H. Major, ed., *Early Voyages to Terra Australis Now Called Australia*, Hakluyt Society Publications, 1st series, vol. 25 (London, 1859), 165–66.

32. Birch, *History*, 4:266.

33. G. Knight, *Collection of Some Papers Published in the Philosophical Transactions of the Royal Society* (1758; repr. London, 1969), 17–18.

34. Newman, *Hudson's Bay Company*, 149; Nordenskiöld, *Periplus*, 49 n. 2; Christy, *North-West Passage*, 279.

35. Logbook "Heemskerck" (VOC), 22 Nov. 1642, ARA, 1.11.01.01/121; logbook "Vrouwe Elisabeth" (VOC), 28 Feb. 1766, NSM, cat.I/313; H. Carrington, ed., *The Discovery of Tahiti: A Journal of the Second Voyage of H.M.S. Dolphin round the world . . .*, Hakluyt Society Publications, 2d series, vol. 98 (London, 1948), 16, 102–6; Millet Dechales, *Art de Naviger*, 99.

36. C. H. Cotter, "The Early History of Ship Magnetism: The Airy-Scoresby Controversy," *Ann. Sci.* 34 (1977): 589–90; A. E. Fanning, *Steady As She Goes: A History of the Compass Department of the Admiralty* (London, 1986), xxii–iv, 421–24; Davids, *Zeevaartkunde*, 53 n. 11; A. R. T. Jonkers, "The Fading Tradition: The Declining Use of Magnetic Declination in Dutch Pelagic Navigation," in *Frutta di Mare: Evolution and Revolution in the Maritime World in the Nineteenth and Twentieth Centuries*, ed. P. C. van Royen, L. R. Fisher, and D. M. Williams (Amsterdam, 1998), 54–57.

37. Iron found in the binnacle in logbook "Propatria" (WHA), 27 June 1784, GADord, 156/3; warnings against deviating iron for instance in Holwarda, *Friesche Sterrekonst*, 240; A. de Graaf, *De Gehele Mathesis of Wiskonst, Herstelt in Zijn Natuurlijke Gedaante* (Amsterdam, 1694), 161; K. de Vries, *Schatkamer ofte Konst der Stierlieden . . .* (Amsterdam, 1702), 80; Y. Valois, *La Science et la Pratique du Pilotage . . .* (Bordeaux, 1735), 160; E. Floryn, *Schatkamer ofte Kunst der Stuurlieden, Inhoudende de Arithmetica of Rekenkunde . . .* (Amsterdam, 1786), 92–93.

38. Remarks on a minimum distance between (steering) compasses in Reael, *Observatien*, 15, 17–18; D. R. van Nierop, *Nieroper Schatkamer, Waermee dat de Kunst der Stuerluyden . . . Gebruickt Kan Worden* (Amsterdam, 1676), 59; Hellingwerf, *Stuurmanskunst*, 67; Harris, *Treatise of Navigation*, 191; iron parts of the pumps were blamed in logbook "Banda" (VOC), 1 Mar. 1636, ARA, 1.04.02/1121; iron rudder supports causing trouble in logbook "Beschermer" (ADM), 22 Jan. 1738, ARA, 1.01.47.17/43; deviation by iron gratings in logbook "Catherine" (EIC), 25 June 1710, IOBL, L/MAR/B 115 B.

39. C. H. Gietermaker, *'t Vergulde Licht der Zeevaart; ofte Konst der Stuurluyden*, 3d ed. (Amsterdam, 1677), 107; Graaf, *Grootte Zeevaert*, 48; Holwarda, *Friesche Sterrekonst*, 347; C. J. Vooght, *De Zeemans Weghwyser* (Amsterdam, 1706), 93.

40. Harrison, *Idea Longitudinis*, 41; Fanning, *Steady As She Goes*, xxii; Hitchins and May, *Lodestone to Gyro-Compass*, 52; Boxer, "Portuguese Roteiros," 172, 175–76; Fontoura da Costa, *Ciência Náutica*, 73; Balmer, *Beiträge*, 203.

41. Logbook "Chandos" (EIC), 6 Sept. 1690, IOBL, L/MAR/A 79; H. C. Freiesleben, *Geschichte der Navigation* (Wiesbaden, 1978), 44.

42. Logbook "Philibert" (CDI), 23–24 Feb. 1734, ANP, MAR 4JJ/114; "Mer des Indes," ANP, MAR 3JJ/330 no. 31; Pietersz, *Stuurmanskunst*, 126; J. Mudbey, RS, MM vol. 7 no. 102 fol. 13, compared binnacle, gangway, forecastle, and poop and generally "found a difference of 3° and sometimes more."

43. Churchman, *Magnetic Atlas*, 4th ed., 50.

44. Markham, *John Davis*, 197.

45. Christy, *North-West Passage*, 309, 316; other references in T. Goueye, *Observations Physiques et Mathematiques pour Servir a l'Histoire Naturelle . . .* (Paris, 1688), 89; A. van Berlicom, *In de Natuirliicke Dingen Aen te Mercken . . . de Oorsaecken van de Bewegingen . . .* (Rotterdam, 1656), 174.

46. A. E. E. McKenzie, *The Major Achievements of Science*, vol. 1 (Cambridge, 1960), 14; Marcorini, *History of Science*, 218; E. G. R. Taylor, "Five Centuries of Dead Reckoning," *J. Inst. Nav.* 3 (1950): 280–81; Taylor, *Haven-Finding Art*, 143–44; E. G. R. Taylor and

M. W. Richey, *The Geometrical Seaman: A Book of Early Nautical Instruments* (London, 1962), 21; Walker, *Treatise of Magnetism*, 193, 195.

47. A. Cabeliau, *Reken-Konst van de Groote See-Vaert* . . . (Amsterdam, 1617), preface; examples of compass points (rather than degrees) allowed in logbook "Venus" (CDI), 14 Oct. 1728, ANP, MAR 4JJ/62; logbook "Brandenbuerg" (WIC), 7 Aug. 1730, MMPH, MSS; logbook "Duke of Argyle" (MER), 30 Sept. 1750, NMM, LOG/M/46; logbook "Patriot" (WHA), 18 July 1787, GAS, MSS K 47/1.

48. Waters, *Art of Navigation*, 25, 155; Fanning, *Steady As She Goes*, xviii; Hellmann, *Rara Magnetica*, 103–4; Thompson, *William Gilbert*, 3; Balmer, *Beiträge*, 102–6; Gilbert, *De Magnete*, 249–50; A. Wolkenhauer, "Beiträge zur Geschichte der Kartographie und Nautik des 15. bis 17. Jahrhunderts," *Acta Cartographica* 13 (1972): 412, 438–42, 458–60; A. Wolkenhauer, "Der Schiffskompas in 16. Jahrhundert und die Ausgleichung der Magnetischen Deklination," *Das Rechte Fundament der Seefahrt: Deutsche Beiträge zur Geschichte der Navigation*, ed. W. Köberer (Hamburg, 1982); De Graaf, *Grootte Zeevaert*, 20, 44–45; Holwarda, *Friesche Sterrekonst*, 262–65; Nautonier, *Mecometrie de l'Eymant*, 15.

49. T. Rundall, ed., *Narratives of Voyages towards the North-West in Search of a Passage to Cathai and India 1496 to 1631* . . . , Hakluyt Society Publications, 1st series, vol. 5 (London, 1849), 253–54; W. Borough, *A Discourse of the Variation of the Compas, or Magneticall Needle* (London, 1596), 158–59; R. Polter, *The Pathway to Perfect Sayling* (London, 1605), C2v; Markham, *Baffin*, 44–45.

50. Flamsteed to Pepys, 21 Apr. 1697, in Forbes, Murdin, and Willmoth, eds., *Correspondence of John Flamsteed*, 2:633.

51. Williams, *Mariner's Compass Compleated*, v–vi; Foster, *Downton*, 46.

52. Taylor, *Tudor and Stuart England*, 30; E. G. R. Taylor, "Jean Rotz and the Marine Chart 1542," *J. Inst. Nav.* 7 (1954): 139; Wolkenhauer, *Beiträge*, 450–51, 458, 461; P. Herigonus, *Cursus Mathematici* (Paris, 1644), 403–4; Riccioli, *Geographiae*, 475; BNP, FR MSS 22046 fol. 127r-v.

53. Coignet, *Instruction*, 10, 12; logbooks "Amsterdam" and "Vriesland" (preVOC), 14 July 1598, ARA, 1.04.01/46; logbook "Mauritius" (preVOC), 25 June 1598, ARA, 1.04.01/43; logbook "Oranjeboom" (VOC), 12 Oct. 1616, ARA, 1.04.02/1061 fol. 235v; Metius, *Geographische Onderwysinghe*, 44; logbook "Wapen van Delft" (ADM), 20–21 Sept. 1623, ARA, SG 1.05.01/9290; Lastman, *Schatkamer*, 44; Graaf, *Grootte Zeevaert*, 20; E. de Decker, *Practyck vande Groote Zee-Vaert* (Rotterdam, 1659), 68; Anhaltin, *Slot en Sleutel*, 47; J. Pietersz, *Sardammer Schatkamer of 't Geopende Slot der Groote Zeevaert* (Amsterdam, 1699), 74; S. Pietersz, *Stuermans Schoole, in Welcke de Navigatie* . . . *Bequamelijck Voorgestelt en Geleert Wert* (Amsterdam, 1659), 74; O. F. de Groot and C. J. Vooght, *De Zeemans Wegh-Wyser* (Amsterdam, 1684), 89; logbook "Jonge Tibbe" (MER), 25 June 1762, FSM, Bib J/46.

54. Thompson, *William Gilbert*, 3; Balmer, *Beiträge*, 107; Taylor, *Bourne*, 86–87, 212; Gellibrand, *Variation*, 5; Markham, *Sarmiento*, 29–30; Hellmann, *Rara Magnetica*, 104; C. Anthonisz, *The Safeguard of Saylers, or Great Rutter* . . . , trans. R. Norman (1590; London, 1632), 70; J. Tapp, *The Seamens Kalender or Ephemerides of the Sunne* . . . , 6th ed. (London, 1617), G4r; Norwood, *Seamen's Practice*, 107–8; J. Tapp, H. Phillippes, and H. Bond, *The Sea-mans Kalender* . . . (London, 1654), 104; Millet Dechales, *Art de Naviger*, 116; D. Newhouse, *The Whole Art of Navigation in Five Books* (London, 1701), 62; Norman, *Newe Attractiue*, 104; De Groot and Vooght, *Zeemans Wegh-Wyser*, 89–92.

55. Atkinson, *Art of Navigation* (1707), 261–62.

56. Davids, *Zeewezen en Wetenschap*, 110; S. Sturmy, *The Mariner's Magazine; or Sturmy's Mathematical and Practical Arts* (London, 1669), 6, 68; H. Gellibrand, *An Epitome of Navigation*, ed. E. Speidell (London, 1698), 54; VOC resolution 16 Apr. 1655, ARA, 1.04.02./103 fol. 604, 606.

57. Rundall, *Voyages Towards the North-West*, 250, 253; Asher, *Henry Hudson*, 93.

58. Christy, *North-West Passage*, 114, 279.

59. Logbook "Royalle" (CDI), 6 Aug. 1682, ANP, MAR 4JJ/90; Denys, *Art de Naviger*, 30, 122, 190; Millet Dechales, *Art de Naviger*, 116; Berthelot, *Traité de la Navigation* (Marseille, 1701), 91–92; other references in logbook "Jeux" (CDI, 1689), ANP, MAR 4JJ/90; logbook "President" (CDI, 1687–88) ANP, MAR 4JJ/93.

60. Logbook "Fridericus Quartus" (AK, 1735), RAK, VA 14/759c; logbook "Cron Printzen" (AK, 1734–35), RAK, VA 14/758; logbook "Anna Sophia" (SE, 1729–31), RAK, 85.b 1+2/359; logbook "West Vlieland" (SE, 1728–29), RAK, 85.b 1+2/353; logbook "Morianen" (SE, 1728), RAK, 85.b 1+2/351; logbooks "Christiansborg" (GK, 1725–27) and "Haabet" (GK, 1724–25), RAK, 85.b 1+2/348b; logbook "Grev Laurvig" (SE, 1725–27), RAK, 85.b 1+2/348d.

61. Logbook "Witte Lam" (MER, 1707–8), ANP, MAR 4JJ/144; logbook "Middelburgs Welvaert" (MCC, 1721–23), RAZ, 20/765; logbook "Gouderak" (ADM, 1737–38), ARA, 1.01.46/2318; logbook "Propatria" (WHA, 1784), GADord, 156/3.

62. Early references in logbook "Mauritius" (preVOC), 6 June 1598, ARA, 1.04.01/43; logbook "Overijsel" (preVOC), 7 Mar. 1600, ARA, 1.04.01/109; logbook "Zutphen" (VOC), 14 June 1632, ARA, 1.04.02/1105; later examples in logbook "Noortbeek" (VOC), 27 Jan. 1718, RAZ, MSS 33.1/184; logbook "Zeepaard" (ADM), 19 Jan. 1785, MMPH MSS; logbook "Pollux" (ADM, 1785), ARA, 1.01.46/1212, preface; examples in navigation manuals: Lastman, *Schatkamer*, 53; De Graaf, *Grootte Zeevaert*, 31; Pietersz, *Stuermans Schoole*, 74; D. R. van Nierop, *Onderwijs der Zeevaert, en Ander Wercken* (Amsterdam, 1670), 46; Nierop, *Nieroper Schatkamer*, 68–69; De Graaf, *Gehele Mathesis*, 164; Hellingwerf, *Stuurmanskunst*, 60–62; G. Maartensz, *Schoole der Stuurluyden* (Hoorn, 1701), 89–90; J. A. van Dam, *De Nieuwe Hoornse schatkamer ofte Konst der Zeevaart* (Amsterdam, 1720), 107; A. Erzey, *Korte Grond-Beginzelen der Navigatie ofte Stuurmanskonst* (Amsterdam, 1777), 50.

63. Examples in logbook "Sparendam" (VOC), 6 Apr. 1672, ARA, VOC 1.04.02/5053; logbook "Fowey" (RN), 23 June 1700, PRO, ADM 52/36; logbook "Nieuwe Hoop" (MCC), 2 July 1776, RAZ, 20/853; logbook "Vis" (MCC), 23 July 1778, RAZ, 20/1139.

64. Examples in logbook "Utrecht" (ADM), 1 Dec. 1783, ARA, 1.10.11.02/41; logbook "Bellona" (ADM), 8 May, 19 May, 9 Sept. 1789, MMPH, MSS; a step size of two and a half degrees was ordered in VOC sailing directions issued from 1769, such as ARA, 1.04.02/3250 fol. 475.

65. Logbook "Amitié" (FAD), 14 Aug. 1730, ANP, MAR 4JJ/34; logbook "Kemphaan" (ADM), 26, 29 July 1782, ARA, 1.01.46/1198. This peculiar practice will be explored in the last chapter.

66. Example in logbook "Royal Duke" (EIC), 7 July 1756, L/MAR/B 614 C: "No sight of the sun for az.th or amp. lately, but I alow a point variation."

S E V E N : Plotting the Third Coordinate

1. Barlowe, *Navigator's Supply*, dedication.
2. Christy, *North-West Passage*, 636.
3. W. Barr and G. Williams, *Voyages to Hudson Bay in Search of a Northwest Passage 1741-47*, 2 vols., Hakluyt Society Publications, 2d series, vols. 177 and 181 (London 1994, 1995), 1:86.
4. Dutch logbook instructions ARA, 1.04.02/5019 (VOC, after 1746), nos. 11–12; GAA, PA 231/223 (ADM, 1749); ARA, 1.01.46/801a (ADM 1755), no. 11; the RN format description in NMM, MSS THO/15B-1762 (1787) fols. 1–10; a French example in Valois, *Pratique du Pilotage*, 222–24; Danish instructions in RAK, AK 169, 173, 174, 184, 188, 193, 194b.
5. Stapel and Boetzelaer, *Oostindische Compagnie*, bk. 1, 1:414 app. 5; a later example in ARA, 1.04.02/5026 (VOC, 1793); French logbook processing in ANP, MAR 4JJ/44, ANP MAR 4JJ/74.
6. "Instruments Divers," ANP, MAR G/100 fol. 190; similar intent in de Chabert, "Observations of Compass Variation" (1780s), ANP, MAR 2JJ/38 no. 2; G. Delisle, notebooks 14–15, ANP, MAR 2JJ/59.
7. "Mer des Indes," ANP, MAR 3JJ/336 no. 11.
8. J. Malham, *The Naval Gazetteer; or Seaman's Complete Guide . . .* (London, 1795), vii–viii; Anhaltin, *Slot en Sleutel*, 56.
9. Nordenskiöld, *Periplus*, 76–79, 101-7, 148; A. Lindsay, *A Rutter of the Scottish Seas*, ed. A. B. Taylor, NMM Maritime Monographs and Reports 44 (1540; London, 1980), 5, 23–24; A. C. Crombie, *Science, Optics, and Music in Medieval and Early Modern Thought* (London, 1990), 87; Parry, *Age of Reconnaissance*, 85–87; D. W. Waters, *The Rutters of the Sea: The Sailing Directions of Pierre Garcie* (New Haven, Conn., 1967), 23; Morais e Sousa, *Sciencia Nautica*, 205-8; Albuquerque, *Contribuicao*, 11; J. C. M. Warnsinck, "Reysgheschrift van de Navigatieën der Portugaloyers," *Itinerario. Voyage ofte Schipvaart van Jan Huyghen van Linschoten naer Oost ofte Portugaels Indiën, 1579-1592*, vol. 4, Werken der Linschoten Vereniging 43 (The Hague, 1939), xix–xliv.
10. De Figueiredo, *Hidrographia Exame de Pilotos*, 16v; logbook "Baron de Breteuil" (FAD), 18 Nov. 1703, ANP, MAR 4JJ/47; G. de Glos, *Epitome de la Navigation Pratique* (Honfleur, 1675), 69–98; Fournier, *Hydrographie*, 123; references to Dutch sailing directions in "Sciences et Arts de la Marine" BN, MSS NAF 9490 fol. 13 (1634), and logbook "Postillon" (FAD, 1698), ANP, MAR 4JJ/46; "Mer des Indes," ANP, MAR 3JJ/329 no. 7; Dutch sailing directions in possession of the EIC in IOBL, L/MAR/A 74, 63–65 (1678), and BODL, Rawlinson papers A 334 fol. 1v, 22–41; Spanish sources in English possession in BL, Sloane MSS 2292 (after 1595); BL, Add. MSS 28140 fol. 34v–47v (1718), and BL, Add. MSS 19297 (after 1735).
11. C. H. Cotter, "A Brief History of Sailing Directions," *J. Inst. Nav.* 36 (1983): 258; K. N. Chaudhuri, "The English EIC's Shipping (ca. 1660-1760)," in *Ships, Sailors, and Spices: East-India Companies and Their Shipping in the Sixteenth, Seventeenth, and Eighteenth Centuries*, ed. J. R. Bruijn and F. S. Gaastra (Amsterdam, 1993), 108; "Instructions for the East-Indies with Useful Observations" (1767), NMM, MS NVP/18 fol. 66, 71; d'Après de Mannevillette, *Directions for Navigating from the Channel to the East Indies . . .* (London, 1769); J. Diston, *The Seaman's Guide, Chiefly the Experience of the Author . . .* (London, 1778).

12. ARA, 1.04.02/1432 (1678); ARA, 1.04.02/2970 (1760) fol. 123; ARA, 1.02.04/8481 (1766–68).

13. Logbook "America" (EIC), 27 Dec. 1695, IOBL, L/MAR/A 72; "Zeloan" is Ceylon (Sri Lanka).

14. Parmentier, "Van Linschoten," 156–59, 161–63, 165–67; G. F. Warner, ed., *The Voyage of Robert Dudley . . . to the West Indies, 1594-1595,* Hakluyt Society Publications, 2d series, vol. 3 (London, 1899), 20–21; J. Dunmore, trans./ed., *The Expedition of the "St. Jean-Baptiste" to the Pacific 1769-70 . . . ,* Hakluyt Society Publications, 2d series, vol. 158 (London, 1981), 232; many colorful examples in P. C. Fenton, "The Navigator as Natural Historian," *Mariner's Mirror* 79 no. 1 (1993): 44–57; logbook "Kingfisher" (RN), 23 June 1704 and 6 June 1705, PRO, ADM 52/200; "St. Helena pigeons" also in logbook "Paramore" (RN), 10 Mar. 1700, BL, Add. MSS 30368; seagulls and large magnetic declination off Madagascar in logbook "Sparendam" (VOC), 14 Apr. 1671, ARA, 1.04.02/1280; an "abundance of silv.r birds and a seal" near the Cape of Good Hope in logbook "Bridgewater" (EIC), 22 May 1720, IOBL, L/MAR/B 42 A.

15. K. J. Lohmann and C. M. F. Lohmann, "Sea Turtle Navigation and the Detection of Geomagnetic Field Features," *J. Inst. Nav.* 51 (1998): 19–20; the animals combine values for inclination and total magnetic intensity to form a nonrectangular positional grid.

16. BL, Add. MSS 19889 fols. 130–31; BL, Sloane MSS 2279 fol. 100.

17. General associations of magnetic declination with coasts in Taylor, *Bourne,* ch. 23; Norwood, *Seaman's Practice,* 108; Millet Dechales, *Art de Naviger,* 100; Dassié, *Routier,* 7–8, 10–11, 24; Le Monnier, "Instruments Divers" (1780), ANP, MAR G/100 fol. 183; specific locations linked with declination in logbook "Utrecht" (preVOC, 1598-1600), ARA, 1.04.01/62; sailing instruction "Nieuw Amsterdam" (VOC, 1636), UBL, BPL 127 E fol. 3; Dassié, *Routier,* 68–71; logbook "Susanna" (EIC, 1683), IOBL, L/MAR/A 72; ANP, MAR 2JJ/59 no. 35; BL, Sloane MSS 3296 (1685) fol. 8; "Recueil de Divers Iourneaux et Remarques de Navigation" (1720s), ANP, MAR 4JJ/74 no. 9; "Route pour le Voyage du Perou en Chine et Retour en France" (1728), ANP, MAR 3JJ/323 no. 4; "Instruction sur la Navigation de l'Isle de France de Diego Garcia" (after 1731), ANP, MAR 3JJ/335 no. 7; "Remarks of the Ship Walpole" (1759), NMM, MSS WEL/45, d'Après de Mannevillette, *Navigation,* 30, 37–39, 211, 219.

18. Carpenter, *Geographie Delineated,* 2:27; Norwood, *Seaman's Practice,* preface; an example of searching for the island St. Helena in logbook "Montagu" (EIC), 22–23 May 1745, IOBL, L/MAR/B 552 J; looking for St. Paul in logbook "Kent" (EIC), 29–30 June 1786, IOBL, L/MAR/B 41 C.

19. Logbook "Royalle" (CDI, 1682–83), ANP, MAR 4JJ/90 fol. 14, 39; logbook "Royal Duke" (EIC), 11 Nov. 1752, IOBL, L/MAR/B 614; "Achin Head" was the name of the north tip of Sumatra; A. Struick, *Verhandeling over de Zeevaartkunde* (Rotterdam, 1768), 308.

20. Logbook "Hawke" (RN), 28–29 Mar. 1669, PRO, ADM 52/1270; $lat.d$ = latitude, ob = observation, $ob.d$ = observed.

21. Logbook "Sirenne" (CDI), 7 May 1720, ANP, MAR 4JJ/90.

22. BODL, Rawlinson MSS A 191 no. 67 fol. 212; "Melville Papers," NLS, MSS 1070 fol. 91r–v; examples of a magnetic coordinate in logbook "Jonas" (EIC), July 1621 (sailing direction), IOBL, L/MAR/A 34; logbook "Navarre" (FAD, 1670-71), ANP, MAR B⁴/4

fol. 320; logbook "Swallow" (RN), 10 Sept. 1708, PRO, ADM 52/278; logbook "Gloucester" (RN), 5–7 Mar. 1741, PRO, ADM 52/602.

23. ANP, MAR 3JJ/330 no. 22 (1757?).

24. ANP, MAR 3JJ/336 no. 12; ANP, MAR 3JJ/323 no. 4.

25. "Horlogerie de la Marine," ANP, MAR G/98 fol. 300 par. 7.

26. Logbook "Jewel" (EIC), 30 Aug. 1637, IOBL, L/MAR/A 61: 1 degree 6 minutes; logbook "Grev Laurvig" (SE), 3 Feb. 1726, RAK, SE 348d: 7 degrees; logbook "Paquebot nr. 4" (FAD), 18–19 July 1788, ANP, MAR 4JJ/109: 10 degrees; logbook "Unie" (VOC), 3 Apr. 1790, NSM, cat.I/36: 6 degrees; logbook "Frankendaal" (WHA), 8 Sept. 1789, NSM, cat.I/266 (WHA): 2 degrees.

27. M. Cortés, *The Arte of Navigation . . .* , trans. R. Eden (1557; London, 1615), 141; an early table in logbook "Wapen van Delft" (ADM, 1623–24), ARA, SG 1.05.01/9290.

28. Nierop, *Nieroper Schatkamer*, 61, 291; Berthelot, *Traité de la Navigation*, 91–92; F. Dassié, *Le Pilote Expert* (Havre de Grace, 1720), 110; Steenstra, *Openbaare Lessen*, 13, 16, 40–41; Pietersz, *Stuurmanskunst*, 40–41, 84–85.

29. Logbook "Anson" (EIC), 22 Dec. 1751, IOBL, L/MAR/B 549 C.

30. Harrison, *Idea Longitudinis*, 42–43; Meynier, *Mémoire sur le Sujet du Prix Proposé par l'Académie Royale des Sciences en l'Année 1729 . . .* (Paris, 1732), xiv; Bouguer, *Nouveau Traité de Navigation*, 90–91; old logbooks as information source in logbook "Duc de Bourbon" (CDI), 10 May 1724, ANP, MAR 4JJ/69 and logbook "Kempthorne" (EIC), 7 May 1685, BL, Sloane MSS 3670; oral tradition in the extract of logbook "Arc en Ciel" (FAD), 27 Dec. 1687, ANP, MAR 4JJ/11; Danish occurrences are discussed in Gøbel, *Danish Asiatic Company's Voyages*, 45.

31. Herigonus, *Cursus Mathematici*, 422–26; Graaf, *Grootte Zeevaert*, 31–32, 48; Decker, *Practyck*, 73–75; Nierop, *Nieroper Schatkamer*, 61; Vooght, *Zeemans Weghwyser*, 93–98; Pietersz, *Sardammer Schatkamer*, 231; Harrison, *Idea Longitudinis*, 37–38.

32. R. Dudley, *Arcano del Mare*, 2 vols., 2d ed. (Florence, 1661); Waters, *Art of Navigation*, 159–61, 528–29; M. D. Gernez, "L'Invention de la Boussole et les Débuts de la Cartographie Nautique," *Communications et Mémoires, Académie de Marine*, new series, 5/23 (Paris, 1947), 11; Crone, Dijksterhuis, and Forbes, *Principal Works*, 411; Wright, *Certaine Errors*, contained a small chart with plotted values of declination; Clerc, *Histoire Naturelle*, vol. 5 contains 362 pages of tabulated data, followed by the seven charts.

33. Santa Cruz, *Book of Longitudes*, 16–18; Albuquerque, *Contribuicao*, 12; Thrower, *Voyages of Edmond Halley*, 57; A. H. Robinson, *Early Thematic Mapping in the History of Cartography* (Chicago, 1982), 84.

34. Fontoura da Costa, *Ciência Náutica*, 68, 72; Morais e Sousa, *Sciencia Nautica*, 203; Balmer, *Beiträge*, 182; Robinson, *Thematic Mapping*, 46; Kircher, *Magnes*, 360 gave extensive instructions on how to make an isogonic chart; the author even boasted he would have gladly produced one himself for the interested reader, had the cost, the very hurried printing, and several other reasons not prevented this.

35. Albuquerque, *Curso*, 234, 245–49; Hellmann, *Ältesten Karten*, 18.

36. BODL, Rigaud MSS 37 fol. 74; Robinson, *Thematic Mapping*, 84; Cook, *Halley*, 282.

37. Thrower, *Voyages of Edmond Halley*, 56–58, 61; Hellmann, *Ältesten Karten*, 5–6, 10; Ronan, *Halley*, 177, 180.

38. Crone, Dijksterhuis, and Forbes, *Principal Works*, 412; Thrower, *Voyages of Edmond Halley*, 59–60, 368–70.

39. "Examen de la Carte de Halley," BNP, MSS NAF 10764 fols. 17–18; E. Halley, "Some Remarks on the Variations of the Magnetical Compass . . . ," *Phil. Trans.* 29 no. 341 (1715): 166–67.

40. W. Mountaine and J. Dodson, *An Account of the Methods Used to Describe Lines on Dr Halley's Chart . . . about the Year 1744 . . .* (London, 1755), 7–8; Taylor, *Haven-Finding Art*, 240; Taylor, *Hanoverian England*, 132–33.

41. Mountaine and Dodson, *Account . . . 1744*, 4–9; W. Mountaine and J. Dodson, *An Account of the Methods Used to Describe Lines on Dr Halley's Chart . . . about the Year 1756 . . .* (London, 1758), 4; Halley's 1700 world chart and Mountaine and Dodson's 1744 version were copied by Bouguer in 1753 and 1769, and by Scarella in 1759; P. Bouguer, *Nouveau Traité de Navigation, Contenant la Théorie et la Pratique du Pilotage*, ed. de la Caille (Paris, 1769), back pages; Scarella, *De Magnete*, 1:back page.

42. Mountaine and Dodson, *Account . . . 1756*, 5; W. Mountaine and J. Dodson, "A Letter . . . Concerning the Variation of the Magnetic Needle . . . ," *Phil. Trans.* vol. 50 pt. 1 (1757): 330–49; Mountaine and Dodson's original dataset is now lost, possibly forever.

43. Mountaine to the RS, 13 Apr. 1758, RS, LP 3 fol. 306.

44. Dulague, *Leçons de Navigation*, 2d ed. (Rouen, 1775), 184; S. G. Franco, *Historia del Arte y Ciencia de Navegar*, 2 vols. (Madrid, 1947), 63; Marguet, *Histoire de la Longitude*, 66; Martinez-Hidalgo y Teran, *Historia*, 102; Taylor, *Haven-Finding Art*, 241; Taylor, *Hanoverian England*, 28; ANP, MAP 6JJ/1–4.

45. Dunn, *Navigator's Guide*, 10.

46. Thrower, *Voyages of Edmond Halley*, 365, app. D, on the mariner: "He is not to expect that we should descend to all the particularities necessary for the coaster, our scale not permitting it"; ANP, MAP 6JJ/30 no. 5 (1794) legend (left side): "But it is to be observed that in both the charts published by Mess.rs Mountaine and Dodson, those gentlemen have made use of Dr. Halley's original chart with very little alteration, either in the situation or bearing of places. The improvement of geography had nevertheless, in fifty years, made as considerable a progress as the change of the variation."

47. Gietermaker, *Licht der Zeevaart*, 87; Bézout, *Cours de Mathématiques*, 244; J. O. Vaillant, *Verhandeling van de Stuurmanskonst* (Amsterdam, 1784), 3; Lasalle, *Cours d'Hydrographie ou de Navigation*, 2 vols. (London, 1787), 185–86.

48. Steenstra, *Openbaare Lessen*, 45; Walker, *Treatise of Magnetism*, 45; Bouguer, *Nouveau Traité de Navigation*, 315.

49. Compare "Instruction Papers for East Indiaman Diane" (1768), ARA, 1.04.02/4826 with "Improved Direction to Sail from the Cape of Good Hope to the Island Ceylon in all Seasons" (5 Sept. 1746), ARA, 1.10.69/420, nearly a literal copy, except for the updated values of magnetic declination.

50. Logbook "Fendant" (FAD), 23, 26, 27 Apr. 1782, BNP, MSS NAF 9449.

51. "Magneticks," RS, CP 9 (2) no. 2 fols. 195–96; ibid. no. 3 fols. 200–201.

52. Moore, *New Systeme*, 195, 257–58; Harrison, *Idea Longitudinis*, 34–36; logbook examples in logbook "Jonah" (EIC), 14 June 1631, IOBL, L/MAR/A 57: "in the lattitude of 31d [S] if you sayle E. or E.b.S., euery 30 leagues you do same, you lessen a degree of variation"; logbook "Concord" (EIC), 24 May 1660, IOBL, L/MAR/A 68.

53. "Moyen . . . pour . . . Déterminer la Longitude . . . par l'Observation de la Déclinaison de l'Aiguille . . . ," ANP, MAR 3JJ/35 no. 6.

54. "Mer des Indes," ANP, MAR 3JJ/330 no. 31 fol. 36; Thrower, *Voyages of Edmond Halley*, 59.

CONCLUSION: Quantifying Geomagnetic Navigation

1. R. N. Bhatacharya and E. C. Waymire, *Stochastic Processes with Applications* (New York, 1990), ch. 1.

2. A. Jackson, A. R. T. Jonkers, and M. Walker, "Four Centuries of Geomagnetic Secular Variation from Historical Records," *Phil. Trans.* A 358 no. 1768 (2000): 971–76.

3. Knapton and Knapton, *Atlas Maritimis et Commercialis*, 162.

4. "Mer des Indes" (1765), ANP, MAR 3JJ/330 no. 31, fol. 29, 37.

5. "Zeilage Ordre om . . . van Nederland . . . naar Straat Sunda te Zeilen" (19 Sept. 1783), RAZ, 255/244 fol. 6.

6. Examples of foreign sailing directions and logbooks among French naval documents include ANP, MAR 2JJ/59 no. 25; ibid. 2JJ/95 no. 48; ibid. 3JJ/329 no. 5; ibid. 3JJ/330 no. 16; ibid. 3JJ/333 no. 5, 9; ibid. 3JJ/336 no. 5, 14; ibid. 3JJ/410 no. 1, 4, 8, 16.

7. Consisting of 152 English, 177 French, 1,658 Dutch, and 1,286 Danish corrections.

8. Examples include A. Bichon (VOC, 1690–98): 10 logs in MMPH, MSS; S. Vis (VOC, 1660–70): 12 logs in ANP, MAR 4JJ/144; M. Geslin (CDI, 1773–85): 11 logs in ANP, MAR 4JJ/144 E; R. Mennes (ADM, 1785–89): 10 logs in ARA, 1.01.46/1229, 1233; R. Tiddeman (RN, 1733–62): 12 logs in NMM, TID/2–11; H. Pieters (MER, 1766–73): 12 logs in FSM, Bib J/42, 47; S. Hoogerzeyl (WHA, 1769–98): 14 logs in GADord, 156/1–10, 12, 14, 15.

9. Harris, *Treatise of Navigation*, 217.

10. ARA, 1.10.11.01/8 (1760s); ARA, 1.04.02/4826 (1769); ARA, 1.04.02/3250 (1769) fol. 475–76, 478, 480; RAZ, 255/244; ARA, 1.04.02/4952 no. 46 (1783) fol. 12.

Glossary

agonic (line). A line connecting all points with zero magnetic declination, on which a compass would point true north.

amplitude observation. Determining magnetic declination by comparing the Sun's magnetic and true direction on the horizon, at sunrise and/or sunset.

animism. Metaphysical doctrine that attributes soul or conscious life to inanimate natural phenomena.

antipodality. The state of being in diametrical opposition.

Arctic Circle. A parallel at 23°30′ colatitude, the projected path of the celestial north pole on the Earth's surface during diurnal rotation.

atomism. The philosophy that assumes all matter to consist of a limited variety of tiny, indivisible particles.

aurora. Colorful display in high latitudes, caused by a shower of charged particles from the Sun that collide with oxygen and nitrogen atoms in the atmosphere.

average. See mean.

axial dipole. Dipole aligned with the planet's rotation axis.

azimuth. Horizontal arc, measured from the local meridian.

azimuth observation. Determining magnetic declination by comparing the Sun's magnetic and true direction above the horizon.

celestial pole. Either of the two points in the sky around which the stars appear to revolve.

colatitude. Arc measure of distance from the geographic north pole in a meridional plane.

compass point. $^{1}/_{32}$ part of a circle (11°15′).

convection. Physical transfer of heat in a viscous medium, due to temperature differences.

core. The innermost part of the Earth, made predominantly of iron; the outer part is in a liquid state, the inner core is solid.

Core-Mantle Boundary (CMB). The interface between the liquid outer core and the stony mantle, at a depth of approximately 2,900 km.

core spot. A localized combination of high and reverse magnetic flux, due to the expulsion of a magnetic field beyond the CMB.

corpuscularism. Various mechanistic philosophies that explained natural phenomena by the interaction of tiny particles.

crust. The outermost solid part of the Earth, containing the continental and oceanic plates.

crustal field. The magnetic field caused by magnetized crustal rocks.

dead-reckoning. Incremental positional estimates based on recorded heading and speed/distance run.

declination. See magnetic declination.

declination allowance. Compensation for local geomagnetic declination.

deviation. Needle deflection due to the retained magnetic field of nearby iron, in man-made objects or (crustal) rocks.

dip. See magnetic inclination.

dip circle. See inclinometer.

dip pole. See magnetic pole.

dipole. The simplest magnetic field, with one north and one south pole.

disjointed dipole. A dipole whose poles are not in diametrical opposition.

diurnal variation. Variations in the geomagnetic field in the course of a day, due to the external field.

doldrums. Area where two opposing trade winds meet.

drift. Ship's displacement due to currents.

ecliptic. The Earth's orbital plane around the Sun.

ecliptic pole. Either of the two points at 90 degrees altitude in a plane perpendicular to the ecliptic.

ephemeris. Tabulated predictions of the movement of celestial bodies at regular intervals.

fixed-needle compass. Compass in which a given amount of magnetic declination is compensated by permanently fixing the needle at an angle with the north of the card.

flux. Flow; *see also* magnetic flux.

flux line. A line that is tangent to the flux everywhere.

geographic equator. The great circle at 90 degrees distance from both geographic poles.

geographic north. The direction along the shortest arc of a great circle intersecting the location of the observer and the geographic north pole.

geographic poles. The two points where all meridians intersect, at 90 degrees north and south latitude; the two points where the rotation axis intersects the Earth's surface.

geomagnetic equator. The intersection of the Earth's surface with a plane perpendicular to the dipole and through the center.

geomagnetic poles. The two points where the dipole axis cuts the Earth's surface.

great circle. The intersection of a plane with a sphere that cuts the latter in two equal halves; *see also* small circle.

great-circle degree (spherical degree). Degree as measured along a great circle.

gyrocompass. A nonmagnetic compass, using electric gyroscopes to maintain orientation.

inclinometer. Instrument to measure magnetic inclination, using a vertical magnetic needle pivoting on a horizontal axis.

isoclinic (line). Closed curve connecting all points with the same magnetic inclination.

isogonic (line). Closed curve connecting all points with the same magnetic declination.

isogonic chart/map. Graphical representation on a coordinate grid of isogonics (usually at regular intervals).

latitude. Arc measure of distance from the geographic equator in a meridional plane.

latitude-sailing. See parallel sailing.

leeway. A ship's sideways displacement due to wind-force acting on the hull, masts, and rigging.

lodestone/loadstone. Any iron-rich, magnetic mineral (such as magnetite and hematite).

log. A float attached to a light piece of rope with knots at fixed intervals, used to measure vessel speed.

longitude. Arc measure of distance from a prime meridian in a plane parallel to the equator.

loxodrome. Line of constant heading, cutting all meridians at the same angle.

lunar(-distance) method. Technique to calculate longitude from the position of the Moon relative to various stars.

magnetic declination. The horizontal angle between local magnetic and geographic north (or, between the compass needle and the local meridian).

magnetic equator. The largest closed curve connecting points where geomagnetic field lines run parallel to the surface.

magnetic flux. The constant "flow" of magnetic force that constitutes a magnetic field; the concentration of field lines in a given volume.

magnetic inclination. The vertical angle between the full geomagnetic vector and its horizontal component.

magnetic intensity. Magnetic field strength.

magnetic meridian. The intersection of the vertical plane through the local geomagnetic vector with the Earth's surface.

magnetic north. The direction along the local horizontal magnetic vector.

magnetic philosophy. Any philosophy explaining natural phenomena with the aid of magnetism.

magnetic pole. Any location where the geomagnetic field lines run perpendicular to the Earth's surface.

magnetic storm. Sudden intensification of the solar wind, resulting in disturbances in the Earth's magnetic field.

mantle. The stony shell enveloping the Earth's core.

mean. The sum of all observations divided by their number.

median. The middle observation in a sorted series.

meridian. The shortest line on the Earth's surface that connects the geographical poles.

meridian distance. Physical distance from a meridian along a parallel.

mimesis. Direct imitation or representation of an aspect of nature, by possession of (some of) the same attributes.

misfit. The quantified disagreement between model and observations.

model resolution. The critical scale below which no features can be discerned.

monsoon. Seasonally alternating winds in low latitudes due to temperature differences between sea and land.

multipole. A system exceeding dipole complexity.

non-antipodality. *See* disjointed dipole.

northeasting/northwesting. Eastward/westward magnetic declination.

northern lights. *See* aurora.

octant. Instrument to measure arc in the vertical plane, up to 90 degrees.

parallax. An object's apparent change in position when viewed from different points.

parallel. A small circle of constant latitude (that is, parallel to the equator).

parallel sailing. Sailing due east or west on the latitude of the intended destination.

precessing dipole. A tilted dipole whose geomagnetic poles orbit the geographical poles over time.

Prime Mover/Primum Mobile. In Aristotelian cosmology, the ninth sphere that conveys movement to the eight inner ones that bear the stars and the known planets around the Earth.

quadrupole. A system containing four magnetic poles.

rectifier. Moveable card on top of a regular compass card, turned at will to dynamically compensate for local magnetic declination.

refraction. Light bending with the Earth's curvature, due to differences in air pressure and temperature, causing objects to appear higher in the sky than they actually are.

reverse-flux. Magnetic attraction of opposite polarity from that expected from a purely dipolar field.

rhumb. See loxodrome.

roteiro. (Oceanic) sailing direction.

"running down the latitude." See parallel sailing.

scalar. A quantity with magnitude only.

secular variation. Changes in the geomagnetic field of internal origin on a timescale from years to millennia.

sextant. Instrument to measure arc in the vertical plane, up to 120 degrees.

sextupole. A system containing six magnetic poles.

small circle. The intersection of a plane with a sphere that cuts the latter in two unequal halves; *see also* great circle.

solar wind. The emanation of charged particles from the Sun.

spherical degree. See great-circle degree.

standard deviation. The square root of the variance.

"swinging the ship." Tracking the influence of deviation from ship's iron by tabulating compass behavior on all headings while turning the ship full circle about a vertical axis.

sympathy. Metaphysical doctrine postulating mutual affinity or reciprocal dependence of bodies.

terrella. A small spherical piece of lodestone used to investigate geomagnetism. (Literally "little Earth.")

tilted dipole. A dipole oriented at some angle from the planet's rotation axis.

time-dependent model. A dynamic model spanning a period of time.

"touching the needle." Magnetization of a compass needle by stroking it with a piece of lodestone.

trade winds. Steady westerlies, blowing from ca. 30 degrees north and south latitude toward the equator.

true north. Geographic north.

variance. The sum of the squared residuals from the mean, divided by the number of observations minus one.

variation. See magnetic declination.

vector. A quantity with magnitude and direction.

westward drift. The westward movement of magnetic field anomalies in the Atlantic hemisphere over historical time.

zenith. The celestial point right above the observer, 90 degrees of arc distant in altitude from the horizon.

zodiac pole. See ecliptic pole.

Essay on Sources

This book is an edited version of my Ph.D. thesis *North by Northwest: Seafaring, Science, and the Earth's Magnetic Field (1600–1800)* (Amsterdam, 2000), which contains a bibliographical framework that fully acknowledges the many published and unpublished sources relied upon. Here, only a few general scholarly publications will be mentioned, those that may serve to elaborate the historical background and the main concepts treated in the text. This is followed by brief reviews of the most relevant printed and manuscript primary sources respectively.

SCHOLARLY WORKS

Starting with geophysics, a comprehensive and accessible introduction to many aspects of deep Earth studies is C. M. R. Fowler, *The Solid Earth: An Introduction to Global Geophysics* (Cambridge, 1992). A more recent alternative is provided by W. Lowrie, *Fundamentals of Geophysics* (Cambridge, 1997). Other introductions covering the Earth sciences are G. D. Garland, *Introduction to Geophysics (Mantle, Core, and Crust)* (Philadelphia, 1971); S. Vogel, *Naked Earth: The New Geophysics* (New York, 1995); and G. A. Good, *Sciences of the Earth: An Encyclopedia of Events, People, and Phenomena*, Encyclopedias on the History of Science vol. 3 (New York, 1998).

Works specifically dealing with the Earth's magnetic field include W. D. Parkinson, *Introduction to Geomagnetism* (Edinburgh, 1983); W. H. Campbell's *Earth Magnetism: A Guided Tour through Magnetic Fields* (Burlington, Mass., 2001), and his *Introduction to Geomagnetic Fields* (Cambridge, 1997). Among recent journal articles describing geomagnetic field modeling without extensive mathematics are J. Bloxham and D. Gubbins, "The Evolution of the Earth's Magnetic Field," *Scientific American* 261 (1989): 68–75, and "The Secular Variation of Earth's Magnetic Field," *Nature* 317 (1985): 777–81, by the same two authors. A standard work on this subject is J. A. Jacobs, ed., *Geomagnetism*, 4 vols. (London, 1987), which contains a separate historical section by S. R. C. Malin, "Historical Introduction to Geomagnetism," in vol. 1, pp. 1–48. Studies of the dynamics in the outer core (where the field is generated) have been conducted from the 1950s; a detailed review is given in P. H. Roberts and A. M. Soward, "Dynamo Theory," *Annual Review of Fluid Mechanics* 24 (1992): 459–512. The current research itself figures in A. Jackson, A. R. T. Jonkers, and M. Walker, "Four Centuries of Geomagnetic Secular Variation from Historical Records," *Philosophical Transactions of the Royal Society* A 358 no. 1768 (2000): 957–90.

Moving on to the history of science, a well-written introduction is D. Goodman and C. A. Russell, eds., *The Rise of Scientific Europe, 1500–1800* (Sevenoaks, 1991). For the history of geophysics, see, for instance, R. McCormach, L. Pyenson, and R. S. Turner, eds.,

Historical Studies in the Physical Sciences (Baltimore, Md., 1979). An extensive bibliography on the subject is supplied by S. G. Brush, H. E. Landsberg, and M. Collins, *The History of Geophysics and Meteorology: An Annotated Bibliography,* Bibliographies of the History of Science and Technology vol. 7, Garland Reference Library of the Humanities vol. 421 (New York, 1985).

General histories of magnetism in the context of that of electricity can be found in P. Benjamin, *The Intellectual Rise in Electricity: A History* (New York, 1895); P. Fleury Mottelay, *Bibliographical History of Electricity and Magnetism Chronologically Arranged* (London, 1922); J. Daujat, *Origines et Formation de la Théorie des Phénomènes Électriques et Magnétiques,* Exposés d'Histoire et Philosophie des Sciences 989–91, 3 vols. (Paris, 1945); and in J. L. Heilbron, *Electricity in the Seventeenth and Eighteenth Centuries: A Study of Early Modern Physics* (Berkeley, Calif., 1979). Particular attention to French developments is given in C. S. Gillmor, *Coulomb and the Evolution of Physics and Engineering in Eighteenth-Century France* (Princeton, N.J., 1971).

Magnetic investigations are, moreover, the main subject in A. Still, *Soul of Lodestone: The Background of Magnetical Science* (New York, 1946), and R. W. Home, "'Newtonianism' and the Theory of the Magnet," *History of Science* 15 (1977): 252–66. A sociological approach is taken in several articles by S. Pumfrey; these include "Mechanizing Magnetism in Restoration England: The Decline of Magnetic Philosophy," *Annals of Science* 44 (1987): 1–22; "'O Tempora, O Magnes!': A Sociological Analysis of the Discovery of Secular Magnetic Variation in 1634," *British Journal for the History of Science* 22 (1989): 181–214; "'These Two Hundred Years Not the Like Published as Gellibrand Has Done The Magnete': The Hartlib Circle and Magnetic Philosophy," in *Samuel Hartlib and Universal Reformation: Studies in Intellectual Communication,* ed. M. Greengrass, M. Leslie, and T. Raylor (Cambridge, 1994). Lastly, the early development of the geomagnetic discipline is highlighted in H. Balmer, *Beiträge zur Geschichte der Erkenntnis des Erdmagnetismus,* 3 volumes, Veröffentlichungen der Schweizerischen Geselschaft für die Geschichte der Medizin und Naturwissenschaft 20 (Aarau, 1956). Some of the graphical representations of the Earth's magnetic field have been reviewed in P. Radelet-de Grave, *Les Lignes Magnétiques du XIIIème Siècle au Milieu du XVIIIème Siècle,* Cahiers d'Histoire et de Philosophie des Sciences, new series, no. 1 (Paris, 1981). Regarding more recent developments, consult C. S. Gillmor and J. R. Spreiter, eds., *Discovery of the Magnetosphere,* History of Geophysics vol. 7 (Washington, D.C., 1997).

Comprehensive histories of magnetism tend to reserve little space for practical navigational concerns (such as the longitude problem), while specific attention paid to scientific geomagnetic exploits was, until recently, largely centered on a select company of "great men," such as Gilbert, Plancius, Stevin, and Halley. For works on Gilbert, see S. P. Thompson, *William Gilbert and Terrestrial Magnetism in the Time of Queen Elizabeth: A Discourse* (London, 1903); Fleury Mottelay, *Bibliographical History* (see above); and D. H. D. Roller, *The "De Magnete" of William Gilbert* (Amsterdam, 1959). For Plancius, see J. Keuning, *Petrus Plancius, Theoloog en Geograaf 1552-1622* (Amsterdam, 1946). Stevin is discussed in E. J. Dijksterhuis, *Simon Stevin: Science in the Netherlands around 1600* (The Hague, 1943); E. Crone, E. J. Dijksterhuis, and R. J. Forbes, *The Principal Works of Simon Stevin,* vol. 3 (Amsterdam, 1961); and D. J. Struik, *The Land of Stevin and Huygens: A Sketch of Science and Technology in the Dutch Republic during the Golden Century,* Studies in the History of Modern Science vol. 7 (Dordrecht, 1981). For Halley, see A. Armitage, *Edmond Halley* (London, 1966); N. J. W. Thrower, ed., *The Three Voy-*

ages of Edmond Halley in the Paramore, 1698-1701, 2 vols., Hakluyt Society Publications, 2d series, vol. 156 (London, 1980); and the recent biography by A. H. Cook, *Edmond Halley: Charting the Heavens and the Seas* (Oxford, 1998).

Regarding navigation in general, a starting point could be periodicals such as the *Journal of the Institute of Navigation* and *Mariner's Mirror,* as well as the bibliographies given in standard works such as the forthcoming *Encyclopedia of Maritime History,* edited by J. B. Hattendorf (New York, in press). For French navigation, consult, for instance, publications by F. Marguet, *Histoire Générale de la Navigation du XVe au XXe Siècle* (Paris, 1931), and F. Russo, "L'Hydrographie en France aux XVI et XVIII Siècles: Écoles et Ouvrages d'Enseignement," in *Enseignement et Diffusion des Sciences en France au XVIII siècle,* ed. R. Taton (Paris, 1964). Dutch bibliographies can be found in works by E. Crone, in particular in "Het Vinden van de Weg over Zee van Praktijk tot Wetenschap," *Mededelingen van de Koninklijke Vlaamsche Academie voor Wetenschappen, Letteren en Schone Kunsten van België* 25 no. 9 (Brussels, 1963). Other works from this author in this context include "De Vondst op Nova Zembla: een Hernieuwd Onderzoek der Navigatie Instrumenten," *Bulletin Rijksmuseum* 14 no. 2 (1966): 71-85; and "De Zestiende-Eeuwse Zeeman als Kartograaf," *KNAG Geografisch Tijdschrift* 1 (1967): 173-84. The general Dutch maritime encyclopedia *Maritieme Geschiedenis der Nederlanden* moreover features three contributions on navigation by G. Schilder and W. F. J. Mörzer Bruyns, entitled respectively "Zeekaarten en Navigatie," vol. 1 (Bussum, 1976); "Navigatie," vol. 2 (Bussum, 1977); and "Navigatie," vol. 3 (Bussum, 1977). The links between Dutch science and practical navigation are thoroughly explored in C. A. Davids, *Zeewezen en Wetenschap: de Wetenschap en de Ontwikkeling van de Navigatietechniek in Nederland tussen 1585 en 1815* (Amsterdam, 1986). Finally, for English references, a starting point could be J. B. Hewson, *A History of the Practice of Navigation* (Glasgow, 1951); E. G. R. Taylor, *The Haven-Finding Art* (London, 1956); and the extensive work by D. W. Waters, *The Art of Navigation in England in Elizabethan and Early Stuart Times* (London, 1958).

Concentrating more particularly on the lodestone, the magnetized needle, and the compass, quite a few authors have written on the introduction of these devices. A. Radl, *Magnetstein in der Antike, Quellen und Zusammenhänge* (Stuttgart, 1988), investigated the lodestone in antiquity, and U. Schnall, *Navigation der Wikinger: Nautische Probleme der Wikingerzeit im Spiegel der Schriftlichen Quellen* (Hamburg, 1975), discussed the Viking era. Meanwhile, A. Wolkenhauer published several important contributions: "Beiträge zur Geschichte der Kartographie und Nautik des 15. bis 17. Jahrhunderts," *Acta Cartographica* 13 (1972): 392-498; "Der Nürnberger Kartograph Erhard Etzlaub," *Acta Cartographica* 20 (1975): 504-25; and "Der Schiffskompas in 16. Jahrhundert und die Ausgleichung der Magnetischen Deklination," in *Das Rechte Fundament der Seefahrt: Deutsche Beitrage zur Geschichte der Navigation,* ed. W. Köberer (Hamburg, 1982). The earliest traces of compass use and magnetic declination have been inventoried by L. de Saussure, "L'Origine de la Rose de Vents et l'Invention de la Boussole," *Archives des Sciences Physiques et Naturelles* 3-4 (1923); by E. O. von Lippmann, *Geschichte der Magnetnadel bis zur Erfindung des Kompasses (gegen 1300),* Quellen und Studien zur Geschichte der Naturwissenschaften und der Medizin, vol. 3, pt. 1 (Berlin, 1932); and in publications by H. R. Winter, notably "Seit Wann ist die Missweisung Bekannt?," in *Annalen der Hydrographie und Maritimen Meteorologie,* Zeitschrift für Seefahrt- und Meereskunde, Herausgegeben von der Deutschen Seewarte Hamburg

(Berlin, 1935): 352–63, 416; "The Pseudo-Labrador and the Oblique Meridian," *Imago Mundi* 2 (1937): 61–73; and "Die Erkenntnis der Magnetischen Misweisung," *Forschungen und Fortschritte* 15 (1939): 36–39. French developments are discussed in M. D. Gernez, "L'Invention de la Boussole et les Débuts de la Cartographie Nautique," *Communications et Mémoires*, Académie de Marine, new series, no. 23 (Paris, 1947). A more recent addition to the literature is J. A. Smith, "Precursors to Peregrinus: The Early History of Magnetism and the Mariner's Compass in Europe," *Journal of Medieval History* 18 (1992): 21–74.

The compass within the wider framework of the evolution of instruments has been described, for the fifteenth and sixteenth century, by F. R. Maddison, *Medieval Scientific Instruments and the Development of Navigational Instruments in the XVth and XVIth Centuries,* Agrupamento de Estudos de Cartografia Antiga 30 (Coimbra, 1969); for the seventeenth century by S. Lilley, "The Development of Scientific Instruments in the Seventeenth Century," in *The History of Science: Origins and Results of the Scientific Revolution* (London, 1951), and in the context of scientific instruments by W. D. Hackmann, "Scientific Instruments: Models of Brass and Aids to Discovery," in *The Uses of Experiment: Studies in the Natural Sciences,* ed. D. Gooding, T. Pinch, and S. Schaffer (Cambridge, 1989).

A voluminous discussion of the contents of English navigation manuals from the Elizabethan period (including coverage of compass use) is given by Waters, *The Art of Navigation* (see above). Publications devoted to the history of the compass in general include H. L. Hitchins and W. E. May, *From Lodestone to Gyro-Compass* (London, 1955), and numerous articles by the same W. E. May over the period 1949–1982 in maritime-historical periodicals. A more recent source is A. E. Fanning, *Steady As She Goes: A History of the Compass Department of the Admiralty* (London, 1986). The Dutch case was given substance by Davids, *Zeewezen en Wetenschap* (see above), while a list of Dutch compass makers was recently compiled by S. ter Kuile and W. F. J. Mörzer Bruyns, *Amsterdamse Kompasmakers ca. 1580–ca.1850: Bijdrage tot de Kennis van de Instrumentmakerij in Nederland* (Amsterdam, 1999).

Finally, the problem of determining longitude at sea has received widespread coverage since the time it was recognized as such. Twentieth-century sources include E. G. Forbes, *The Birth of Scientific Navigation: The Solving in the Eighteenth Century of the Problem of Finding Longitude at Sea,* NMM Maritime Monographs and Reports vol. 10 (London, 1974); F. Marguet, *Histoire de la Longitude à la Mer au XVIIIe Siècle en France* (Paris, 1917); D. W. Waters, "Nautical Astronomy and the Problem of Longitude," in *The Uses of Science in the Age of Newton,* ed. J. G. Burke (Berkeley, Calif., 1983); E. Guyot, *Histoire de la Détermination des Longitudes* (Chaux des Fonds, 1955); C. A. Davids, "Finding Longitude at Sea by Magnetic Declination on Dutch East Indiamen, 1596–1795," *The American Neptune* 50 no. 4 (1990): 281–90; and W. J. H. Andrewes, ed., *The Quest for Longitude* (Cambridge, Mass., 1996). Of these, only Davids has paid substantial attention to reliance on geomagnetic hypotheses to solve the issue. As far as sailing directions are concerned, the book by A. E. Nordenskiöld, *Periplus: Essay on the Early History of Charts and Sailing Directions,* trans. F. A. Bather (Stockholm, 1897), still offers the most extensive treatment. The relationship between nautical and scientific endeavors has additionally been investigated by C. H. Cotter, in numerous articles in the *Journal of the Institute of Navigation,* and in his book *A History of Nautical Astronomy* (London, 1986). More information can also be found in D. Howse, ed., *Five Hun-*

dred Years of Nautical Science 1400-1900, Proceedings of the Third International Reunion for the History of Nautical Science and Hydrography (London, 1981).

PUBLISHED PRIMARY SOURCES

A surprisingly large number of original printed works have survived from the studied period. Treatises on (geo)magnetism are most often found in national or university libraries (foremost the British Library, London, and the Bibliothèque Nationale, Paris), and in museums of the history of science. These are also good places to find complete series of historical scientific periodicals of note, such as the *Philosophical Transactions of the Royal Society of London,* and the *Mémoires de l'Académie Royale des Sciences, Paris.* Sea-related sources, such as navigation textbooks, sailing directions, and nautical almanacs, are of course best researched in maritime museums (notably the National Maritime Museum, Greenwich, and the National Shipping Museum, Amsterdam). Other maritime sources (among them dozens of logbooks from famous or very early voyages) have appeared in annotated reprint editions. The two most sizeable series (over one hundred volumes each) are the *Hakluyt Society Publications* (series 1 and 2) and the *Werken der Linschoten Vereniging.*

Some of the oldest tracts involving magnetic observation have been reprinted in a series by G. Hellmann. These include *E. Halley, W. Whiston, J. C. Wilcke, A. von Humboldt, C. Hansteen, Die Ältesten Karten der Isogonen, Isoklinen, Isodynamen, 1701, 1721, 1768, 1804, 1825, 1826: Sieben Karten in Lichtdruck mit einer Einleitung,* Neudrucke von Schriften und Karten über Meteorologie und Erdmagnetismus no. 4 (Berlin 1895, repr. Nendeln 1969); *Rara Magnetica, 1269-1599,* ibid. no. 10 (Berlin 1898, repr. Nendeln 1969); and "Die Anfänge der Magnetischen Beobachtungen," *Acta Cartographica* 6 (1969): 174–98. Translated and facsimile editions of landmark publications include the work of Peregrinus, in P. Radelet-de Grave and D. Speiser, trans./ed., "Le 'De Magnete' de Pierre de Maricourt: Traduction et Commentaire," *Revue d'Histoire des Sciences et Leurs Applications* 28 no. 3 (1975): 193–234; and P. Thomsons, ed., *Petrus Peregrinus: Epistle Concerning the Magnet* (London, 1902). For Gilbert, see W. Gilbert, *De Magnete,* trans. P. Fleury Mottelay (New York, 1958). For a nautical example, see P. de Medina, *Regimiento de Navegacion* (Seville, 1563), Ahora Nuevamente Publicado por el Instituto de España (Madrid, 1964).

Scientific correspondences have sometimes also been published. In the case of Flamsteed, Halley, and Euler, consult the following: E. G. Forbes, L. Murdin, and F. Willmoth, comp./eds., *The Correspondence of John Flamsteed, the First Astronomer Royal,* 2 vols. (Bristol, 1995, 1997); E. F. Macpike, ed., *Correspondence and Papers of Edmond Halley* (London, 1932); A. A. Cournot, ed., *Lettres de L. Euler à une Princesse d'Allemagne sur Divers Sujets de Physique et de Philosophie,* 2 vols. (Paris, 1842). Lastly, the complete manuscript archive of polymath Samuel Hartlib has been transferred to cd-rom; see M. Greengrass and M. P. Leslie, eds., *Samuel Hartlib: The Complete Edition* (Ann Arbor, Mich., 1995).

UNPUBLISHED PRIMARY SOURCES

As the introduction already made clear, the work before you mainly relied upon two types of manuscript sources: ships' logbooks and geomagnetic tracts. The logbooks themselves were supplemented with a host of other relevant maritime-historical material. Most of the research therefore took place in repositories that were either directly related to seafaring (that is, maritime museums), or held sizeable holdings of particular shipping companies (such as national archives). These will briefly pass review per country.

Starting with England, the main maritime holdings of the nation are kept in three places (all in London): the National Maritime Museum, Greenwich (NMM); the Public Record Office, Kew (PRO); and the British Library (BL). The National Maritime Museum is the proud owner of the largest collection of early-modern navy logs (section /LOG and various family papers), consisting of about five thousand captain's logs, and five thousand lieutenant's logs. Most navigational information is usually found in the master's logs, which are kept at the PRO (section /ADM). This is also the resting place of a sizeable part of the Hudson's Bay Company archives, in microfilm copy (section /BH/1).

For the English East India Company, the India Office at the British Library (IOBL) holds about two and a half thousand remaining logbooks up to 1800 (section L/MAR/A and L/MAR/B), as well as a wealth of other company material. These were formerly kept at a separate location in London, but have now joined the other collections at the new site of St. Pancras. This building also houses the Manuscripts Library, which contains a limited number of other logbooks, and additionally owns several longitude schemes and correspondence pertaining thereto. Printed proposals can moreover be found in the regular pre-1975 catalog of printed works.

Vital additional sources are kept in the archives of London's Royal Society (RS), more particularly that institution's Copy Journalbook (/CJB), the Classified Papers (/Cl.P. or /CP), the Letters and Papers (/LP), and the complete series of the *Philosophical Transactions*. In addition to the Society's manuscript collections, its library on the history of science has also proved a valuable aid in research.

Moving on to France, the state's centralized bureaucracy has ensured that almost all material of note has ended up in Paris. Some geomagnetic and longitude tracts are now housed in the Bibliothèque Nationale (BNP); see in particular the sections "French Manuscripts" (/FR) and "New French Acquisitions" (/NAF). This library also possesses a number of navigation manuals from the studied era.

Maritime sources were found in abundance at the Parisian Archives Nationales (ANP). Here are kept most surviving documents of the Compagnie des Indes, as well as vast collections of naval files, including many hundreds of logbooks from both organizations. Note that the logbooks have not been stored separately per organization, but are ordered by destination (in maritime section /MAR/4JJ). Large quantities of hydrographical material are furthermore kept under other numbers of the /MAR/#JJ series. More information pertaining to compasses and their applications can be found in sections /MAR/B^4, /MAR/G, and Colonial Archives /COL/C^2.

In the Netherlands, the huge archive of the Dutch East India Company (the VOC) rivals that of its English counterpart at the IOBL, and is kept at the Algemeen Rijksarchief (ARA, General State Archives), in The Hague (section /1.04.02). But despite centralizing efforts, VOC sources remain somewhat scattered over the country; several municipal archives (of those cities that had a company chamber) also hold records, as do some provincial archives. The southern province of Zeeland is furthermore keeper of the legacy of the Middelburgsche Commercie Compagnie (MCC), a large Dutch slaving company. These documents now reside at the Provincial State Archives in Middelburg (Rijksarchief Zeeland, RAZ).

For Dutch naval matters, the first and best place to look is again the Rijksarchief, The Hague (Admiralties: sections */1.01.46* and */1.01.47*). As with the VOC, naval logs may also be kept among the family papers of their authors (section */1.10*). Furthermore, one of the finest collections of materials related to navigation in the widest sense is the one at the National Shipping Museum, Amsterdam (NSM, catalog */I*); other maritime museums in the Netherlands also tend to have at least a few logbooks and navigation manuals. Information on longitude proposals, on the other hand, is scarce; university libraries (notably Leyden and Amsterdam) provide the best access. Correspondence related to the longitude proposal by Meindert Semeyns is, however, found at the Municipal Archives of Delft (GAD).

Danish relevant material is mainly confined to logbooks, kept at the State Archives (Rigsarkivet), Copenhagen. The Danish East and West Indian Companies have all left sizeable holdings there (section */VA XIV*), as have the Danish Navy (section */85.b.1+2*), and the Greenland Companies (section */140*). A large collection of Danish whaling logs was at the time of investigation not yet properly indexed for public access. Danish sailing directions, geomagnetic hypotheses, or longitude proposals have not been found.

Finally, research in Spain was limited to the General India Archive (Archivo General de Indias, AGI) at Seville, and purely focused on geomagnetic hypotheses and related longitude solutions. Correspondence pertaining thereto resides in the sections */Indiferente* 426–28 and */Filipinas* 340. The main proposals are held in */Patronato* 262. It furthermore deserves to be noted that all AGI inventories and manuscripts are accessible in electronic form (on site), greatly facilitating the search for relevant material.

Index